Early Praise f
The Dregg Dis

"If Choose Your Own Adventure® does a geometry workbook next, I'm in trouble . . ."

— Euclid

"Stones, the abacus, the calculator, and now *The Dregg Disaster*: a natural sequence of math advancements."

— Fibonacci

"How students have made it through Algebra I for so long without *The Dregg Disaster* is an enigma."

— Alan Turing

"There should be no such thing as boring mathematics."

— Edsger Dijkstra

"x = algebra, y = understanding, and *The Dregg Disaster* = x + y!"

— Ada Lovelace

CHOOSE YOUR OWN ADVENTURE®

THE DREGG DISASTER

BY CHRIS MATTHEWS

ILLUSTRATED BY MARÍA PESADO
COVER ILLUSTRATED BY EOIN COVENEY

CHOOSECO
WAITSFIELD, VERMONT

Book design: Peter Holm, Sterling Hill Productions

For information regarding permission, write to:

CHOOSECO
P.O. Box 46
Waitsfield, Vermont 05673
www.cyoa.com

Names: Matthews, Chris, author. | Pesado, Maria, illustrator. | Coveney, Eoin, illustrator.
Title: The Dregg disaster / Chris Matthews ; illustrated by Maria Pesado ; cover illustrated by Eoin Coveney.
Description: Waitsfield, VT : Chooseco, 2022. | 74 b&w illustrations. | Series: Choose your own adventure. | Audience: Ages 12-16. | Summary: This interactive story places the reader in the position of figuring out what the problem is at Dregg Corporation, the biggest employer in town. Every choice requires finding answers to algebraic equations in this book designed for students learning Algebra 1.
Identifiers: ISBN 9781937133931 (softcover)
Subjects: LCSH: Mathematics -- Juvenile literature. | Mathematical recreations -- Juvenile literature. | Mathematics -- Problems, exercises, etc. -- Juvenile literature. | LCGFT: plot-your-own stories. | BISAC: YOUNG ADULT FICTION / Action & Adventure / General. | YOUNG ADULT NONFICTION / Mathematics / Algebra.
Classification: LCC QA95.M38 2022 |793.74 M --dc22

Published simultaneously in the United States and Canada

Printed in the United States

10 9 8 7 6 5 4 3 2 1

Love your local public schools.
Public education needs support now more than ever.

BEWARE and WARNING!

This book is different from other Choose Your Own Adventure books. It's certainly different from other math books. Just like other Choose Your Own Adventure books, you will need to be heroic and use your smarts to reach the best ending. Just like other math books, you will need to solve algebra problems as you move from page to page.

Read the story, make your choice, and solve a math problem. You will need to find the correct solution in order to find the next page in your adventure. Just remember: even though there is one correct solution to most of the problems in this book, there are often lots of great ways to find the right answer. Be creative! If you get stuck, there is Adventurer's Advice in the back of each chapter.

A word of caution before you begin your quest: Be careful that the page that you turn to is located in the correct chapter! This book is split into four chapters, and each one has pages numbered 1-50. Don't move on to the next chapter until the book tells you to.

And don't worry. There are still multiple astounding plot twists, unlikely villains, and plenty of death endings.

Good luck!

CHAPTER 1

As you walk home from soccer practice, there's so much on your mind that you almost go right by it. But the metal clasp catches your eye in the late-day sun. It's a woman's clutch wallet, an old one, lying on the sidewalk. Someone must have dropped it. You pick it up and take a peek inside. You see $35, a couple of old receipts, and a restaurant punch-card for Wong's Deli.

Whoever lost their wallet is one punch away from a free sandwich, you think.

You dig further and find a strange access card. It's shiny and metallic, with a blue, stylized "D." You would know *that* logo anywhere. The logo belongs to the Dregg Corporation, and the card probably does too. Their main research facility is here in town, but they have warehouses and offices all over the world. Dregg makes electronics, chewing gum, laundry detergent, and everything in between. You have probably seen this same logo a thousand times.

The card is small but heavy for its size. And it's not a regular Dregg employee ID, because you'd recognize that. Your mom used to be a product manager there, until her skincare website took off. No, this card is for something else.

You turn it over. There's a small holographic photo of a woman in a lab coat. The name "Doctor Donda LaBella" appears underneath. Below that are the words "IF FOUND PLEASE RETURN TO THE DREGG CORPORATION. $1,000 REWARD."

NOTE TO THE READER: On most pages of this story you will be given a choice, and in order to find the next page in the story, you will need to use your math skills! For this first page, solve the problem below, and when you find a solution, go to the page that matches your answer!

See page 53 of Adventurer's Advice Chapter 1 for help.

You should have arrived here from page 8

It's now or never. You've got to make your move.

You pick up a small rock at your feet and toss it to the other side of the pool. The alligator turns that way and snaps at the rock. While the massive creature is distracted, you hustle across the edge to the metal door.

You turn the handle, but it's locked. However, the metal frame that holds the door is rusted, and badly. You push and shove the door back and forth. Some rusty metal crumbles away. At last, there's a snapping sound, and the door breaks free of its hinges and clatters to the cement floor.

You step over the door and into a well-lit space beyond. As your eyes adjust, you hear a loud rumble. Almost like a . . . train? A train! These underground tunnels have brought you to a subway station!

There are several people boarding a nearby train, but with the sounds of the subway, nobody noticed the noise you made on your way in. You quickly check the subway map painted on the wall, and hurry onto the train. Dregg Tower is only three stops away.

To continue your adventure, turn to page 29.

You should have arrived here from page 24

You rush across to the trash chute. It's made of white canvas with supports spaced out at ten-foot intervals, running all the way to the dumpster below. You pull yourself up, swinging your legs into the opening.

Here goes nothing!

The friction of your jeans against the canvas seems to slow you down just enough so that you're not out of control. But what's at the bottom?

When you're almost to the ground, the chute flattens out. You're able to grab onto one of the supports, and you hold on for dear life. The last thing you need is tetanus from a construction dumpster. The whole chute structure shakes as it absorbs your momentum, then it settles.

Inching forward on your butt, you move to the end of the chute and look down.

The dumpster is empty. It's safe to jump.

Its hollow bottom makes a dull, echoing metal sound as you land. You breathe a sigh of relief that no boards are down here with nails sticking out, or whatever else. You've been lucky so far.

Peeking cautiously over the edge of the big bin, you see a trash truck. It's just pulling out of this alley—and it's about to turn west, in the direction of Dregg Tower.

If you hurry, you could jump on that little metal platform in back and catch a ride.

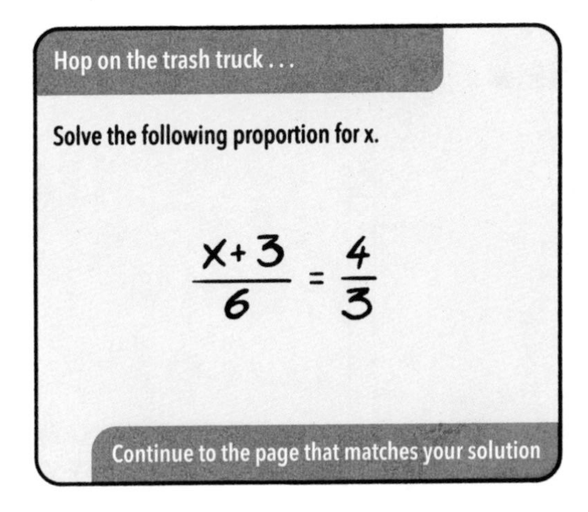

See page 59 of Adventurer's Advice Chapter 1 for help.

...ould have arrived here from page 26

You decide to make a break for it and take the stairs. Still clutching Doctor LaBella's wallet, you clamber down two flights and peek out from the ground-floor stairwell into the street.

That second van is now parked along the road leading to Dregg Tower. You can't just stroll past it—they've surely got a description of you. Plus, the Dregg ID in this wallet seemed to be sending some kind of signal. If you wait here any longer, those goons in suits could come at you from all directions.

An ice cream truck rounds the near corner and comes your way. It's almost dark by now—you could probably dash out and jump onto the back.

You also see, a half block away, the entrance to a subway station, with stairs going down. In the dimming light you *might* be able to get to those stairs undetected. If you can get into the station safely, there are trains every few minutes that stop at Dregg Tower Station.

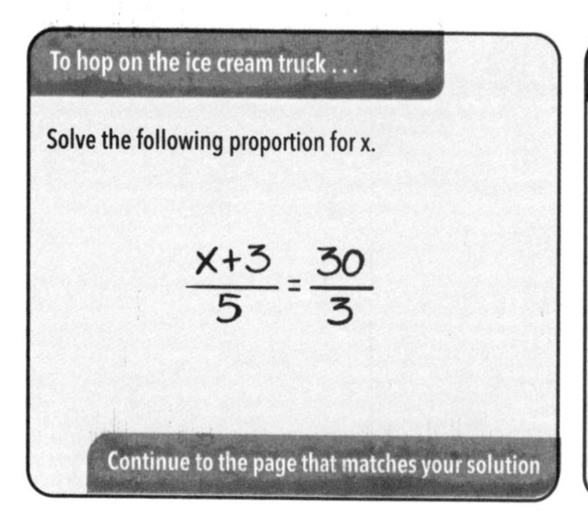

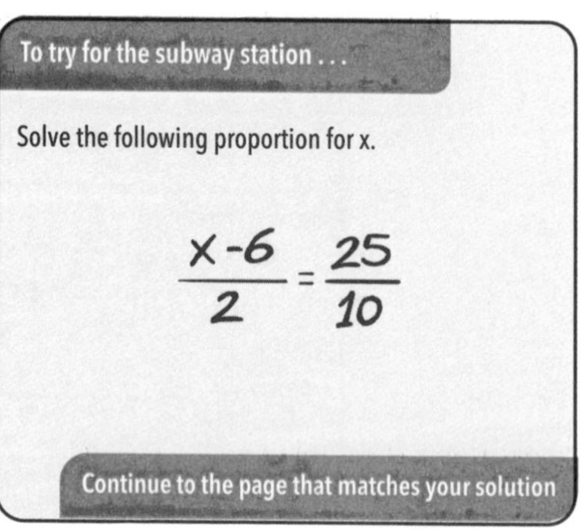

See page 59 of Adventurer's Advice Chapter 1 for help.

You should have arrived here from page 3 OR page 20

It's a cinch to hop on the trash truck's rear platform. As you hold onto the metal railing beside you, the truck rattles and creaks up the street.

Whew. That was close.

Up ahead looms the ominous black glass facade of Dregg Tower, stronghold of the corporation. A block away from it, the trash truck stops at a red light. You hop off onto the street and look around.

You don't see any men in suits coming your way. Are you really going to head straight for the heart of Dregg Corp?

That's right, you think. *Time to find some answers.*

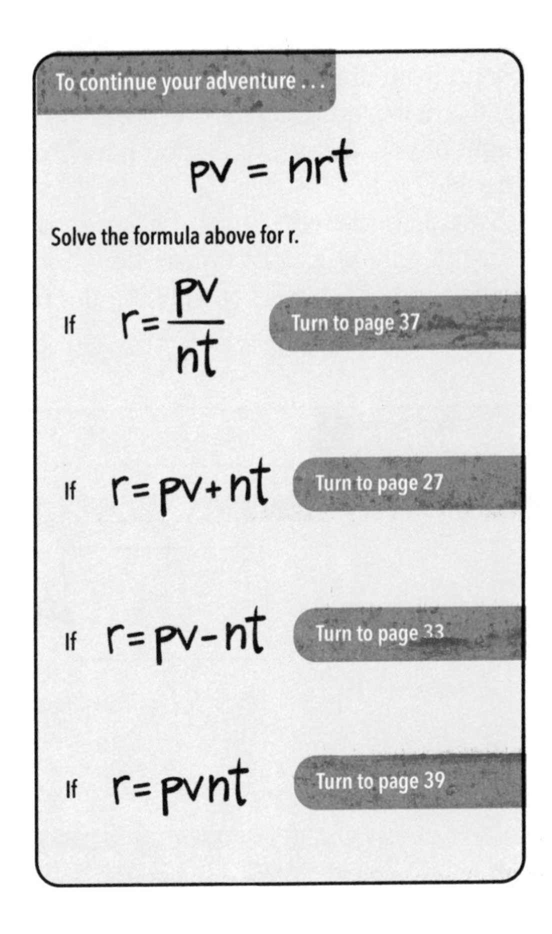

See page 57 of Adventurer's Advice Chapter 1 for help.

You should have arrived here from page 18

The late afternoon is cool, and a breeze makes it cooler. *I can walk fast,* you think, *no problem.* So you set off, Doctor LaBella's wallet in your jacket pocket.

You've gone just a few blocks when—it's funny—you almost *feel* the beeping before you actually hear it. But the sound is definitely getting louder.

Beep . . . beep. It seems to be coming from the wallet.

Puzzled, you pull it out and open it up. It seems like that heavy metallic Dregg ID card is . . . could it be? Yes. Every second or so, it's pulsing. With each pulse, it's also vibrating, and—okay, this is *weird*—with each pulse the raised "D" logo emits a slight flash of eerie blue light.

Looking up and around in confusion, you see that you're no longer alone on the street. Across the street and back a ways, a man in a suit is holding up a small tablet, looking at the screen. Now he looks up, right at you. He lifts a hand from the tablet and says something into his wrist.

Walking quickly, the man starts to cross the street. He's coming toward you, still looking right at you. What is going on here? Could this ID card be sending out a tracking signal?

The guy cups his palm over what must be an earpiece. He listens, nods, and says something into his wrist transmitter.

That's all you need to see. You take off running the other way.

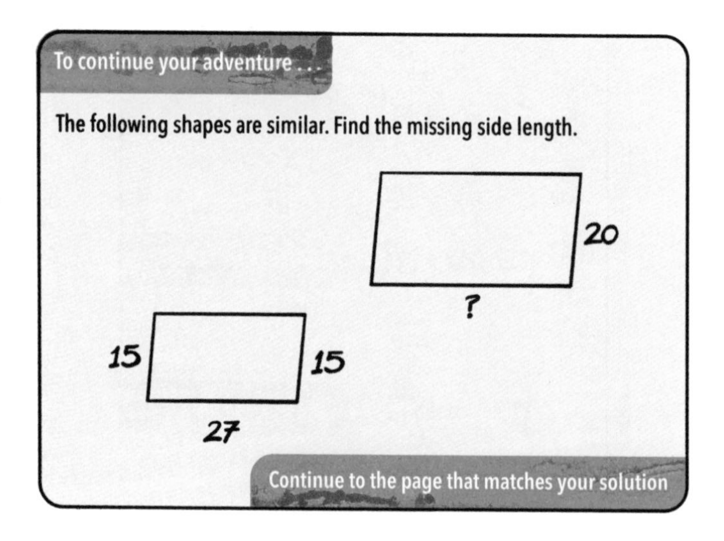

See page 60 of Adventurer's Advice Chapter 1 for help.

You should have arrived here from page 12 OR page 31

You climb up the wet metal rungs carefully. Above your head, the grate is crisscrossed with metal bars. Is it bolted down? Is this a dead end?

Gripping the top rung with one hand, you push hard with the other. The grate is heavy. At first it doesn't budge. Fighting back a feeling of panic, you give everything you've got to one hard, last-hope heave . . .

And the grate comes loose.

Struggling under the weight of the metal grate, you're able at last to push it so that it falls away backward. Grabbing the edge of the opening with both hands, you pull up with all the strength you've got left . . . and heave your upper body out through the opening and onto a hard, black surface.

You pull yourself all the way out and collapse. For at least half a minute, all you can do is lie there and breathe. *Is the reward money worth all this work?*

You have no idea where you are. You lift yourself up on an elbow and look around.

To continue your adventure, turn to page 43.

You should have arrived here from page 44

Slowly, the alligator approaches. Its beady little eyes don't seem to match the rest of its massive frame.

Your eyes dart back and forth as you size up your options.

Beyond this pool, if you can make it, there's a steel door set in the wall of the tunnel. You have no idea if it's locked, or where it would lead. But it's a door. The problem is that you need to cross to the other side of the drain to reach it. Meaning you need to step past the gator.

Up ahead, in another dim pool of light, you can see a metal ladder, like the one you climbed down. It's on the same side of the tunnel and maybe you can get out that way. If this alligator doesn't get you first.

See you later, alligator. Try making it to the steel door . . .

Solve the following proportion for x.

$$\frac{2x+5}{12} = \frac{3}{4}$$

Continue to the page that matches your solution

In a while, crocodile. Try to reach the ladder . . .

Solve the following proportion for x.

$$\frac{1}{5} = \frac{9}{3x+9}$$

Continue to the page that matches your solution

See page 59 of Adventurer's Advice Chapter 1 for help.

You should have arrived here from page 41

A few powerful kicks push you on a strong angle across the current, toward that culvert. You can still hear voices, but they're farther up the river now. It doesn't seem like you've been spotted . . . yet.

There's no grate across the culvert's opening, which is about five feet across. Big enough to easily swim through. You reach the opening and grab for the edge of the big concrete pipe—but something wraps around your leg. Your leg is wrapped up tight, and you feel a spiking panic.

What is it? Some kind of weed? A tentacle?

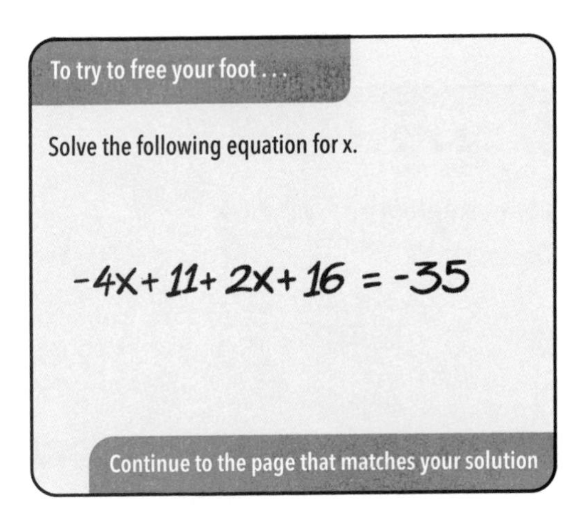

See page 55 of Adventurer's Advice Chapter 1 for help.

You should have arrived here from page 26

You wait here for a few minutes. The concrete floor is cold. You hear a vehicle coming quickly, and you hug the hard floor. The vehicle speeds by, heading down toward the exit. You can only glimpse its wheels and a bottom slice of its body, but you're pretty sure you recognize the silver paneling of a Dregg security van.

Carefully, quickly, you duck over to the stairwell, push open its door, and make your way down the stairs. You don't know what the Dregg goons will do next in their frantic push to get this wallet back—and you don't want to find out.

At street level, you spot a subway station just down the block. You stride as fast as you can toward the station, while also trying your best to act like your rush is just casual.

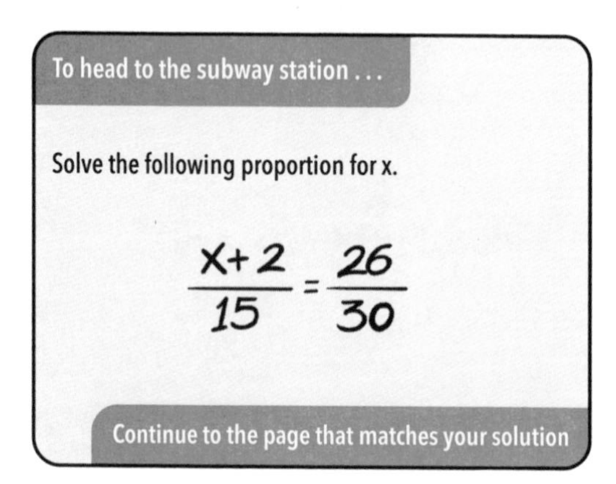

To head to the subway station . . .

Solve the following proportion for x.

$$\frac{x+2}{15} = \frac{26}{30}$$

Continue to the page that matches your solution

See page 59 of Adventurer's Advice Chapter 1 for help.

This is Adams Street Station, you realize as you reach it. You hustle down the stairs in the dimming light. A clock that hangs from the station ceiling says it's after 6:00 PM. Rush hour is mostly over, but there are still people walking by with briefcases and shoulder bags, heading for the trains to go home.

You pluck a dollar bill from Doctor LaBella's wallet and use the money to buy a ticket to Dregg Plaza. That station is just a couple of stops away. You stand on the platform, trying to look casual. After a couple of minutes you spot the headlights from the next train coming down the tunnel. Your train pulls in and eases to a stop. Its doors open with a sighing whoosh, and you hurry in.

Still glancing around nervously, you find a seat by a window, then look out. No men in dark suits come running. The doors slide closed, the train moves off, and you sit back to rest for a few precious seconds.

Through the dark subway tunnel, the train is hurtling toward Dregg Tower.

To continue your adventure, turn to page 29.

You should have arrived here from page 8

The gator lurches your way and your adrenaline surges. Your reaction, luckily, is immediate. Before you even realize that you're moving, you have sprinted through the narrow dry space alongside the ravenous reptile and dashed down the tunnel to safety.

The metal ladder is bolted to the floor of the tunnel and goes straight up to the storm grate. Its metal rungs are a little wet. You grab tight and step up to climb.

To climb up the rungs of the ladder, turn to page 7.

You should have arrived here from page 1

NOTE TO THE READER: Thirteen donuts! A baker's dozen! Great work on navigating this book so far. If you solved the problem on page 1 and got a solution that brought you here, you are off to a great start! Good luck as the choices (and the math problems) get harder!

That settles that! A thousand bucks reward money? Hello gaming laptop.

But how should you return the card and wallet? You could go straight to Dregg Tower and hand it in at Security. Or you could try finding Doctor Donda LaBella directly. She might be grateful enough to give you a bonus. This card is obviously valuable.

Your phone buzzes in your pocket a bunch of times, distracting you. You look at it quickly. One's just a notification from your friend Taylor, who wants you to take your turn on Slime Ninja.

You asleep or something?? Take your turn! Taylor texts.

"Not now!" you say, swiping it away. The other message is an emoji of a noodle bowl from your friend Hazar. Your stomach grumbles. You and Hazar left soccer together, and he's already home to dinner. You don't answer Hazar and put your phone back in your pocket. A thousand dollars is on the line here, and you've got to decide: go to Dregg Tower, or find Doctor LaBella?

NOTE TO THE READER: This time, there is a choice! Decide which path you want to follow and solve the problem that goes with that choice. Your solution will take you to your next page!

If you want to return the wallet and strange ID card to Dregg Tower . . .

Solve the following proportion for x.

$$\frac{x}{15} = \frac{21}{7}$$

Continue to the page that matches your solution

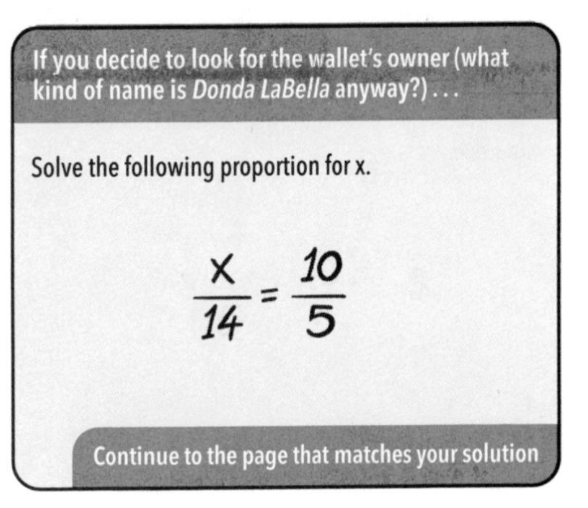

See page 58 of Adventurer's Advice Chapter 1 for help.

You should have arrived here from page 32 OR page 36

You swing your legs into the storm drain opening and quickly climb down the iron ladder into the cement tunnel below.

The storm drain is *just* tall enough for you to walk around, and it extends far to your left and to your right. The darkness in each direction is punctuated by dim light streaming through other storm grates above at street level.

To your right, you hear a loud SPLASH.

What was that?

You can continue down the storm drain toward the sound. That direction seems better lit. Or you can go the other way to avoid *whatever it was* that made that noise.

To head toward the splashing sound . . .

Use order of operations to simplify the following expression.

$$2+3\cdot(4-2)+6^2$$

Continue to the page that matches your solution

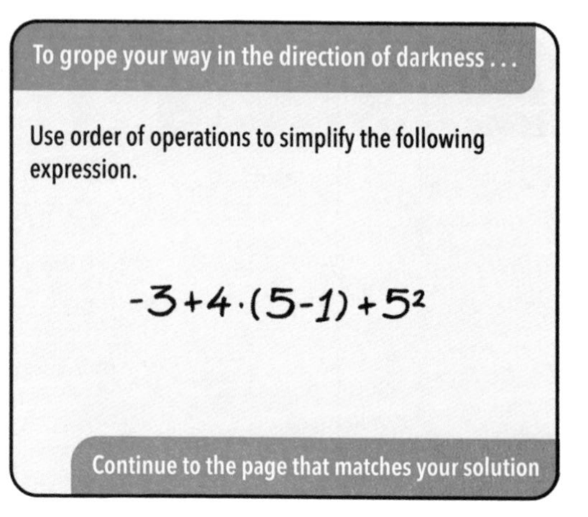

See page 52 of Adventurer's Advice Chapter 1 for help.

You should have arrived here from page 19

You think skinny thoughts and wiggle yourself through the narrow opening. Success! You make it through and move cautiously forward, peering around in the dim light.

Is this a subway station?

It is! You're at the far end of a long concrete platform. Tracks run alongside it, a few feet below the platform. You can see lights ahead and a few scattered people. No one looks your way.

"*Ecchhhhh*," comes a voice from a bench behind a column. Something about this weird, throat-clearing sound feels like it's directed at you. You see a man in old-fashioned clothes sitting on the bench, staring straight ahead.

You don't have time for this, you tell yourself. The tunnel lights up as the train is arriving. But the stranger stands and begins to walk toward you.

"Ex—cuse me?" the man begins. "Ecchhhhhhh." He pauses and clears his throat again.

The train whooshes into the station, and as it does, the man fixes his eyes on you and grabs you by the shoulder. He has very pale blue eyes, with pupils as small as ticks.

"*Tell LaBella I want my sight back*," he commands, so loud you jump backward. The train has stopped and you dash off, leaving the stranger behind on the platform.

You slide through the subway doors just as they are closing, and you take a window seat. You look out through the window, but the man on the platform is gone.

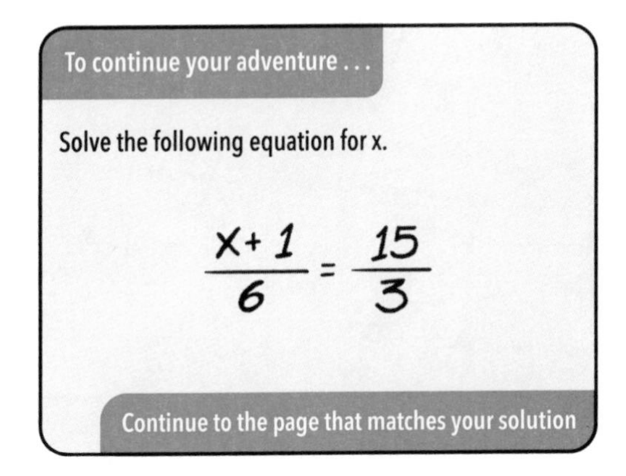

See page 59 of Adventurer's Advice Chapter 1 for help.

You should have arrived here from page 42

You lope along the sidewalk in fast soccer-player strides. Glancing back, you don't see those security guys. Did they give up? Or maybe call for backup?

You turn a corner and peek out again. No men in dark suits and shades. You duck into an alley to catch your breath and think.

Why do those guys want this wallet so badly? The Dregg ID must be an access card with some sort of tracking device, that's why they could follow you. It must have been signaling when it was pulsing and glowing. Who is this Donda LaBella lady?

You dig around in the wallet and find a driver's license with a matching name. There's also a small stack of business cards. Each one has the Dregg logo above these words: *Doctor Donda LaBella, Research Scientist.*

Before you get caught and accused of stealing this wallet, or worse, you need to get to Dregg Tower and find Doctor Donda LaBella.

CHAPTER 1 GOAL: **Reach Dregg Tower with the wallet and access card.**

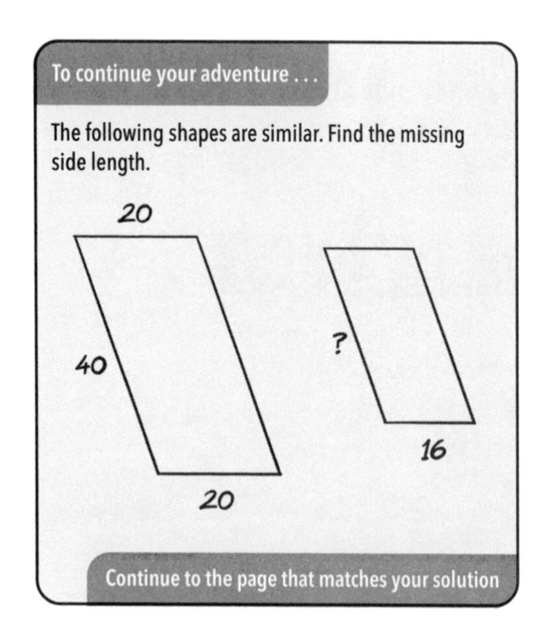

See page 60 of Adventurer's Advice Chapter 1 for help.

You should have arrived here from page 41

You swim for the dock.

Gasping for air and poking your head up every few strokes, you make steady progress toward the dock as the voices slowly fade behind you. Finally you reach it. Holding onto slimy, wet wood, you peek back upriver. In the streetlights on the bridge you can just see a few figures. They're still scanning the water directly below.

Excellent. Your daring escape worked!

Hidden by the dock, you pull yourself out, dripping and heavy with river water. You scramble up onto a dark street next to the river—and seem to be alone. But then a flashlight beam cuts through the darkness.

Uh-oh.

Your sneakers squeak loudly as you turn and run. After what seems like a mile, you can still hear someone following—and they're not far off. Is it the Dregg goons from the van?

Up ahead there's a storm drain. You hurry over, bend low, and stick your fingers into the small rectangular opening. You lift, hard—and it comes loose. Pulling it free, you see a metal ladder going down into . . . into what?

You are standing next to a chain-link fence surrounding a building under construction. You could probably hide somewhere inside, *if* you can make it over that fence.

To start down the ladder into the storm drain, turn to page 35.

To climb the fence and hide in the construction site, turn to page 40.

You should have arrived here from page 28

Dregg Tower is something like two miles away. You could save time, and get home sooner, if you use a little of the cash in the wallet to take a cab.

On the other hand, if you do find Doctor LaBella and she realizes her cash is missing, she might accuse you of stealing it. You could explain that you had to take a cab, but that could get awkward. And you *can* walk to Dregg Tower.

It's up to you.

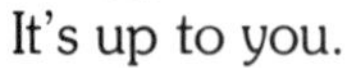

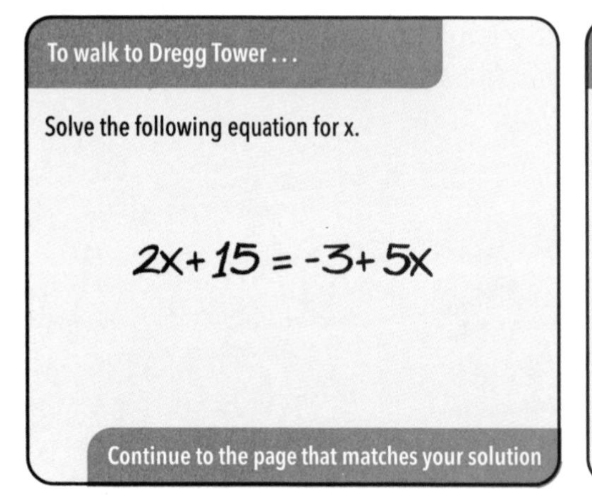

To walk to Dregg Tower . . .

Solve the following equation for x.

$$2x + 15 = -3 + 5x$$

Continue to the page that matches your solution

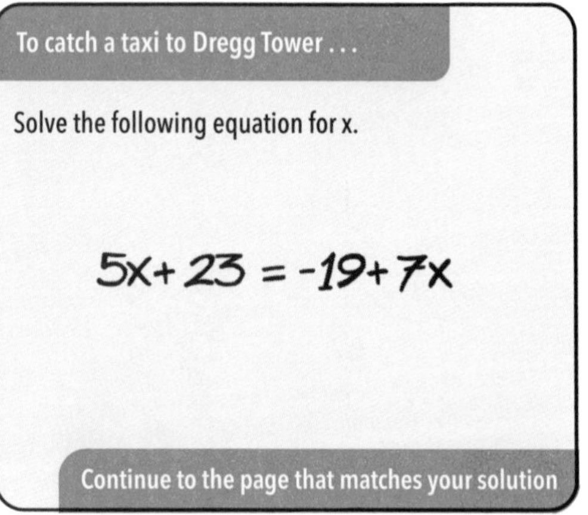

To catch a taxi to Dregg Tower . . .

Solve the following equation for x.

$$5x + 23 = -19 + 7x$$

Continue to the page that matches your solution

See page 56 of Adventurer's Advice Chapter 1 for help.

You should have arrived here from page 30 OR page 40

In the dim light you step carefully down the stairs into the darkness below. This is scary. You can't see a thing. But the deeper you can go down here, the safer you'll be—from whoever was chasing you, at least.

You drag your left hand against the wall while you descend, until you realize your hand is trembling against the rough concrete of the wall. You force yourself to breathe slowly. Hopefully it will help you to calm down.

At last, your feet find level ground. Up ahead, you can see light. You're in an empty concrete room lit by a single overhead bulb.

Something moves at the edge of the room, but you can't quite make out what it is. It's across the room, by a heavy metal grate secured by a padlocked chain.

Uh-oh. You don't want to think about what might be down here with you. You suddenly remember having dinner with your dad at a roadside BBQ restaurant and spotting a giant rat out back by the trash. All of a sudden, rats are all you can think about.

Is this grate the only way out of here? Should you try to get through?

You rush over and pull the grate to test it. The chain has enough give that you can pry the grate open just about twelve inches. Is it enough for you to squeeze through?

The voices up above of whoever is chasing you are louder now. You have no choice. You've got to try.

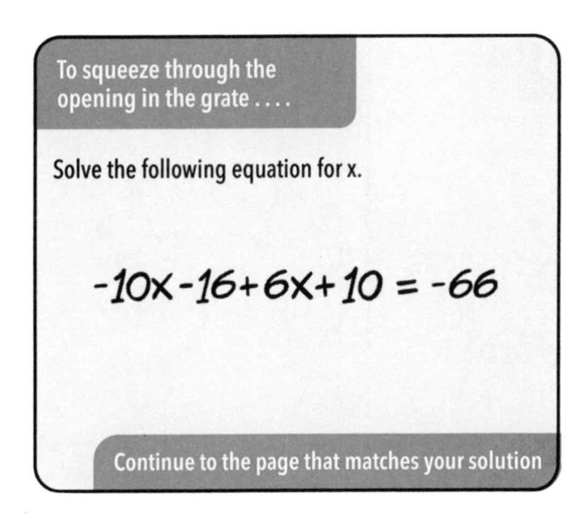

See page 55 of Adventurer's Advice Chapter 1 for help.

You should have arrived here from page 24 OR page 34

Feeling a little like Batman, you grapple your way, hand over hand, down the thick cable. Luckily you and Taylor have been going to the climbing gym by school, and this descent doesn't scare you one bit. And nobody seems to spot you from the alley below.

You jump the last few feet and roll when you hit the ground. Now for the hard part. Where to go?

Peeking out from the alley, you see a garbage truck coming your way. It's headed in the direction of Dregg Tower and no one is standing on the little metal platform that sticks out from the back.

If you're quick, you should be able hop onto that platform. But if you do, whoever is following you might see you.

Looking up the street, you also see a subway station, with stairs leading down into it. The subway stops at Dregg Tower—and by now it's almost dark, so you should make it the half block to those stairs without being spotted.

If you're lucky, that is.

To hop on the back of the trash truck . . .

Solve the following proportion for x.

$$\frac{x+2}{21} = \frac{1}{3}$$

Continue to the page that matches your solution

To make a run for the subway . . .

Solve the following proportion for x.

$$\frac{x-4}{14} = \frac{1}{2}$$

Continue to the page that matches your solution

See page 59 of Adventurer's Advice Chapter 1 for help.

You should have arrived here from page 18

You scroll through your contacts until you see Calloway Cab Co., and call.

The cab must have been right around the corner, because after less than a minute it pulls up, a tan Toyota. You hop in the back seat and give the address.

"Dregg Tower, huh?" says the driver, a woman with short blond hair. "That place gives me the creeps."

After turning onto a busy avenue, the driver weaves through traffic, then glances at her rearview mirror.

"Hey kid," she says, "is somebody following you? I could swear that van back there is on my tail."

You look back. A silver van is behind you, and when your cab shifts lanes, it does too. As it changes lanes, you can just make out the writing on the side:

DREGG CORP. A COMPANY YOU CAN TRUST.

"Um," you ask, "could you maybe go faster? Please?"

"Sure thing, kid!" She floors it.

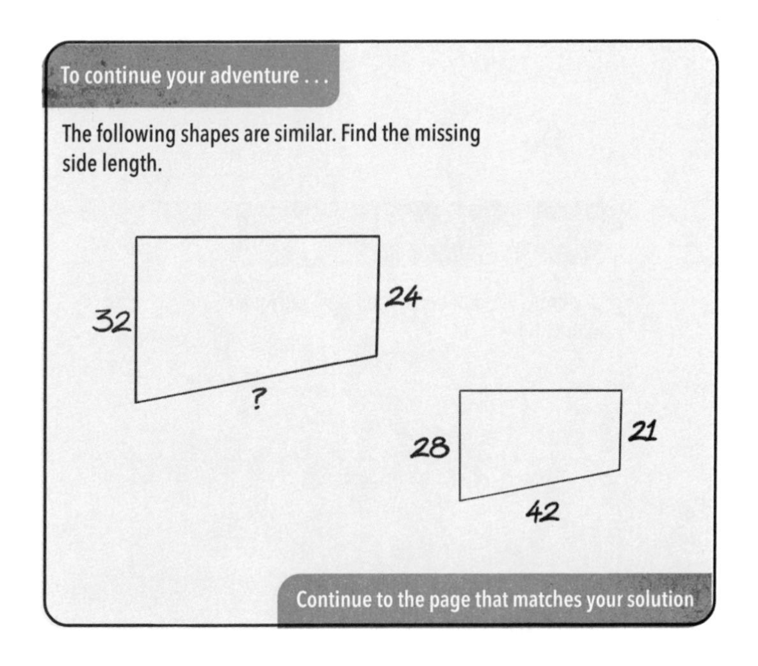

See page 60 of Adventurer's Advice Chapter 1 for help.

You should have arrived here from page 48 OR page 49

"The bridge! Take the bridge!" you cry.

The tires squeal as the woman cranks a hard turn. The pine-tree air freshener hanging from the mirror dances back and forth.

Suddenly at the far end of the bridge another van appears, wheeling around, blocking traffic in both directions. The familiar Dregg logo is painted on its side. Behind you, the first van is careening onto the bridge. You're trapped!

The woman turns to you. "These guys mean business. You better hop out and run for it," she says.

"You're right. Thanks for the help," you say, paying quickly and scrambling out of the car.

Now you're standing on the bridge as men in dark suits and sunglasses weave between the stopped cars from both sides.

You run to the edge of the bridge. It's twenty, maybe twenty-five feet to the water below. You'll have to leap out as far as you can to clear the concrete footers. But right now it doesn't seem like you have any other choice.

"Hey you!" someone yells from way too close. "Stop right there—do not move!"

You pull in a deep breath, close your eyes . . . and jump.

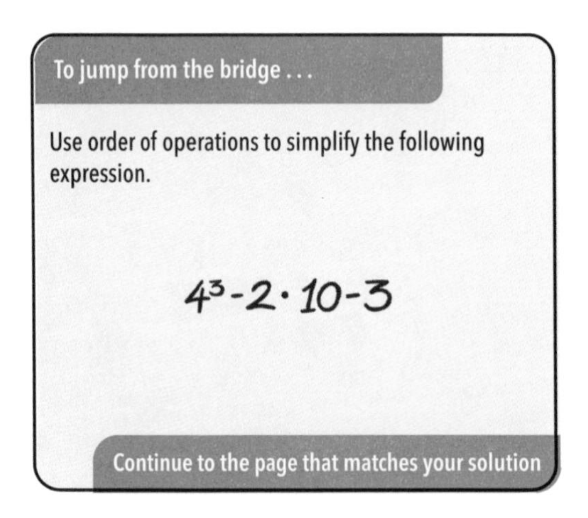

See page 52 of Adventurer's Advice Chapter 1 for help.

You should have arrived here from page 38

What kind of fish has teeth? You step back. As you do, one of the fish leaps out of the water and toward your leg. Its sharp teeth rip through your pants and dig into your leg.

These are piranhas!!

You stumble back, trying to twist and shake the fish off. But you lose your footing on the mossy surface. You fall to the ground, and momentum carries you toward the edge of the pool. You try to grab ahold of anything to stop your slide, but everything that you touch is covered with the same slick moss.

Your slide carries you over the edge and into the underground piranha pool. The last sound you hear is water splashing violently as the hungry fish devour their meal. Their meal is YOU.

The End

You should have arrived here from page 30 OR page 40

The stairs going down are just too dark. You start up the staircase instead.

The handrail is wobbly, but you keep going, past the landing for the skeletal second floor, then the third. Now you're on the roof, and it's wide open up here. From below you hear shouting—and footsteps.

"Yes, in our sights now," you hear someone say, as though they are updating someone over the phone. "Won't be long. No need to worry, it's just a kid."

They're coming up the stairs after you!

You scan the rooftop frantically for an escape in the dimming light. It's too high to jump—and no way can you head back down those stairs.

On one side of the roof, a mini crane arm extends out over the edge, with a pulley that holds a thick cable. A big knot secures the cable to the pulley. You could probably shimmy down the cable—if that knot can hold your weight.

Then you notice a trash chute in the far corner, used by the construction workers when they're here. It's a long white tube, about two feet across, that leads to a dumpster below. You might be able to jump into the trash chute for a ride to earth and evade your pursuers.

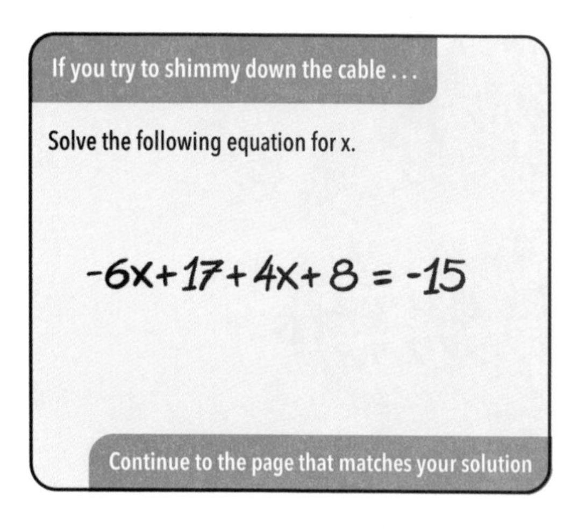

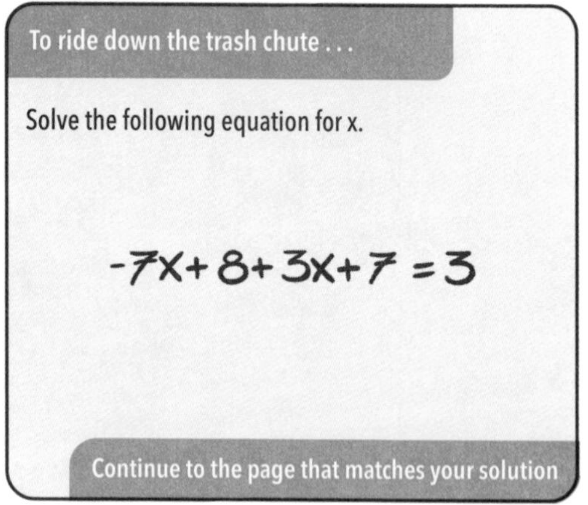

See page 55 of Adventurer's Advice Chapter 1 for help.

You run in front of the approaching car. The driver—a woman with short blond hair—slams on her brakes.

"Help!" you yell. You point to the two men who are now running toward you.

"Hop in!" she answers, leaning to open the passenger door of the tan Toyota. You scramble aboard and are barely inside when she steps on the gas, hurtling forward.

"Who are those guys?" she asks, looking in the rearview mirror. "Why are they chasing you?"

"No idea!" you answer. "They appeared out of nowhere when I found this wallet on the sidewalk."

You pull out the clutch wallet and the glowing metallic card. "Have you ever seen one of these? It's some kind of access card, I think. From Dregg Corporation. It belongs to someone named Doctor Donda LaBella."

"Donda LaBella is a research scientist there. She went missing two days ago. It's been all over the news," the woman says.

She eyes the rearview mirror again. "Your two friends are getting into a van. You don't have much time. Which way do you want me to take you? I'm early to pick up my twins from ballet, so I've got some time."

CHAPTER 1 GOAL: **Reach Dregg Tower with the wallet and access card.**

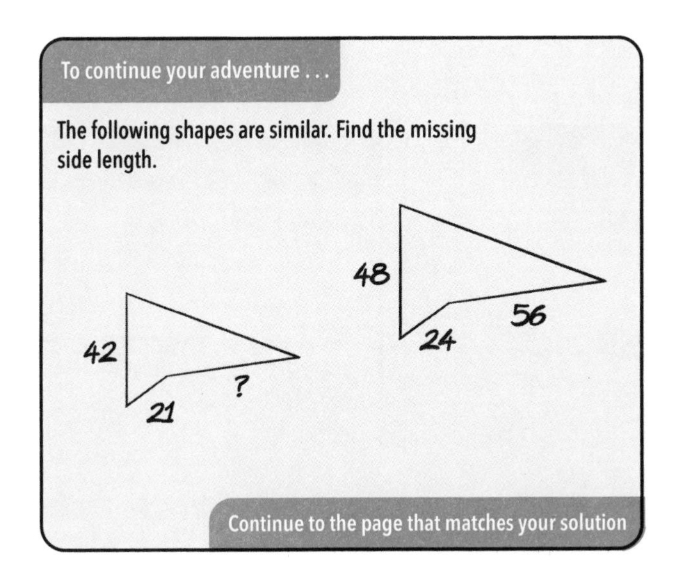

See page 60 of Adventurer's Advice Chapter 1 for help.

You should have arrived here from page 50

"I'll hop out here," you say, paying quickly. "Try to get them to follow you."

She nods and slows just enough for you to leap out. You stoop low and duck behind some cars that are parked to face a concrete wall. "Good luck," she shouts faintly before speeding away.

Tires screech from one level below—they're coming. You lie down flat beneath a car, just in time to see four tires and the bottom of a silver van speed by.

Safe. For now. But there was more than one van . . .

What to do? There's a stairwell nearby, with a big number "2" above the door. You could run down those stairs and make a break for it in the street—but that would mean being exposed. What if another Dregg van is staking out the street?

You could just stay here for a while. They'd have to search hundreds of cars to find you. But if you wait too long, you just might find yourself trapped in this parking garage.

To make a run down the stairwell . . .

Solve the following equation for x.

$$10x + 20 - 15x - 2 = -2$$

Continue to the page that matches your solution

To continue hiding beneath the car . . .

Solve the following equation for x.

$$10x + 10 - 13x + 13 = -7$$

Continue to the page that matches your solution

See page 55 of Adventurer's Advice Chapter 1 for help.

Go back to the previous page and check your work. You should not have arrived here.

You should have arrived here from page 13

You decide you'll take this directly to Doctor Donda LaBella. Why involve security? You can handle it. Maybe something else in the wallet will lead you to her.

There's a driver's license for Donda, and some of her Dregg Corp business cards. Under the name "Doctor Donda LaBella," the cards say, "Research Scientist."

All right, then! A research scientist must earn a good salary—Doctor LaBella might give you a generous reward. Maybe you'll get to see her research lab in Dregg Tower too. You have always liked science, all the way back to Ms. Hiatt's class. Just think of all the crazy experiments you could try and the crazy technology you could buy with *Dregg money*.

You're going to need a little time, so you pull out your phone again and call up your mom.

She picks up after the third ring. "Hey honey. How was practice?" You can hear the radio in the background. That means your dad must be home too.

"I'm fine, Mom. Listen, I forgot something at school that, um, I need for homework. I'm going to run back there, so I might be a little late getting home. But I'm fine," you add again, quickly.

"Do you really have to? Won't the school be closed by now?"

"No, there's a concert later, so school's open, and I really need this, um, book for my project with Hazar. I'll call if I'm going to be more than an hour."

"An *hour*? Let me come pick you up. I'll just get the car and . . ."

"No need, Mom—it's totally fine! I'll see you soon!" You really hate lying to your mom. But it seems easier than explaining the wallet and the card.

It's almost 5:00 PM. With luck, a Dregg research scientist will still be at work. You decide to head over there.

CHAPTER 1 GOAL: **Reach Dregg Tower with the wallet and access card.**

To continue your adventure . . .

Answer the following question. If $x + 8 = 10$, what is $x + 16$?

Continue to the page that matches your solution

See page 51 of Adventurer's Advice Chapter 1 for help.

You should have arrived here from page 2 OR page 11 OR page 15

The subway arrives at Dregg Plaza Station in three short stops. You join the small crowd that files out of the cars, hoping for safety in numbers. In a group, you're harder to spot.

You climb the stairs all together to the ground level. Some of the other passengers are Dregg employees, arriving for the night shift. You recognize their matching green polo shirts—and the badges that hang from blue and yellow lanyards.

The plaza up above ground is surrounded by Dregg office buildings, and in the middle there's a fountain with a statue. You've seen this statue before. It's Levi Dregg, founder and CEO of Dregg Corp, with his arms raised in triumph.

He started this company as a teenager, but nobody has seen him in years.

Beyond Levi's statue is the gleaming dark facade of Dregg Tower.

Well, this is it.

You head toward it.

To make your way to Dregg Tower . . .

Solve the following formula for y.

$$ax+by=c$$

If $y=cx-ab$ Turn to page 33

If $y=\frac{c-ax}{b}$ Turn to page 37

If $y=\frac{c}{b}-ax$ Turn to page 39

If $y=c-b-ax$ Turn to page 27

See page 57 of Adventurer's Advice Chapter 1 for help.

You should have arrived here from page 32 OR page 36

Scaling a chain-link fence is your kind of a challenge—you decide to climb it.

The fence surrounding the construction site is about ten feet tall. But you are in such good shape from soccer, you climb it easily.

High five for team sports.

You drop to the ground on the other side and scope out the building. It's about four stories. Steel girders outline where the walls will go. There's a rough wooden staircase. You jog that way, glancing around nervously. You haven't been spotted . . . yet.

You pass a row of smelly portable toilets and reach the stairs almost immediately.

They go up or down. Not too much light in either direction.

To take the stairs up . . .

Use order of operations to simplify the following expression.

$$-6+2\cdot(9-2)+4^2$$

Continue to the page that matches your solution

To take the stairs down . . .

Use order of operations to simplify the following expression.

$$2\cdot(7-2)+7^2-40$$

Continue to the page that matches your solution

See page 52 of Adventurer's Advice Chapter 1 for help.

You should have arrived here from page 9

You kick *hard*, and your foot comes loose. *Free!*

You're panting from the hard fast swim as you reach the edge of the culvert. You push forward into the concrete tunnel, hoping that nobody from the bridge has spotted you. Either way, you'd better not stop here!

It gets darker farther into the tunnel, but the water is getting more and more shallow. It's becoming difficult to swim without hitting the bottom of the cement tunnel.You stop, sinking to your knees, water rushing by. At least, you *hope* it's just water.

It doesn't smell bad, so there's that.

Dripping water everywhere, you push yourself up to a crouch. You can almost stand. The tunnel is pitch black, but up ahead you see a very dim pool of light. Crouch-walking awkwardly toward it, you make out that the light is coming from a grate. It is set into the top of the tunnel, and you can see a metal ladder leading up to it.

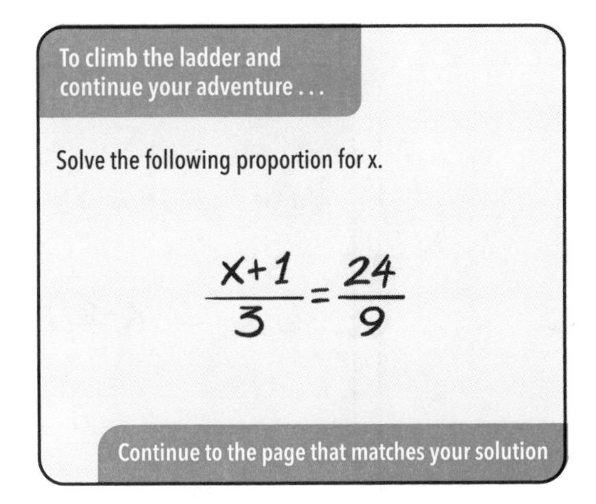

See page 59 of Adventurer's Advice Chapter 1 for help.

You should have arrived here from page 16

You stuff the card back in the wallet, and shove the wallet and your phone into your jacket. You look around at this narrow alley. It's dirty, and it's short—it dead-ends right up ahead, at chain-link fencing surrounding a construction zone. Could you jump it? As you are sizing up the fence, you hear a low whistle coming from the apartment building on the left.

The building has windows from apartment units overlooking the alley, most blocked by air conditioners. Through one window, you can see a pair of eyes looking out at you from a darkened room.

"Hey!" you call out. "Is there another way out of this alley? Or just over that fence?"

"That manhole should lead into a storm drain, you could try that," comes a low voice. "You watch yourself."

You hesitate.

"Who are you?" you ask.

The eyes leave the window, but the voice answers again: "I'm the last thing you should be worrying about. I know who's chasing you, and you better move fast, kid."

To lift the manhole cover over the storm drain . . .

Solve the following equation for x.

$$3(x-10)=12$$

Continue to the page that matches your solution

To hop the fence into the construction site . . .

Solve the following equation for x.

$$2(x-21)=18$$

Continue to the page that matches your solution

See page 54 of Adventurer's Advice Chapter 1 for help.

Go back to the previous page and check your work. You should not have arrived here.

You should have arrived here from page 50

Your driver yanks the steering wheel hard, one way and then the other, as the ramp climbs to the upper levels of the parking garage. Up ahead you can see the evening sky. You've reached the top.

There's a sign: **TOP LEVEL CLOSED FOR CONSTRUCTION**.

Beyond it you can see a cement truck and some piles of bricks and other building materials. But no people. The work crew has most likely left for the night.

With a screech, she pulls to a stop and says, "Better hop out here. Whatever you do now, good luck!" You quickly pay and her vehicle zooms off, leaving you to escape on your own.

You dash past the sign, cement truck, and piles of construction debris, looking for the entrance to the stairs. You spotted a stairwell door on each level of the garage, but they must have just built this new top level. You can't find the stairs anywhere.

Tires squeal as a vehicle, surely the Dregg van, speeds this way. You've got to find some way to escape!

You spot a long orange power cable coiled on the concrete deck. It's the heavy-duty kind, for outdoor use. You might be able to tie that to the wooden railing above the low concrete wall of this deck, then shimmy down the cord.

There's also the cement truck. You could climb inside the mixer and hide out until the coast is clear. You'd have to text your mom you'll miss dinner, but you could say you're having pizza with Taylor.

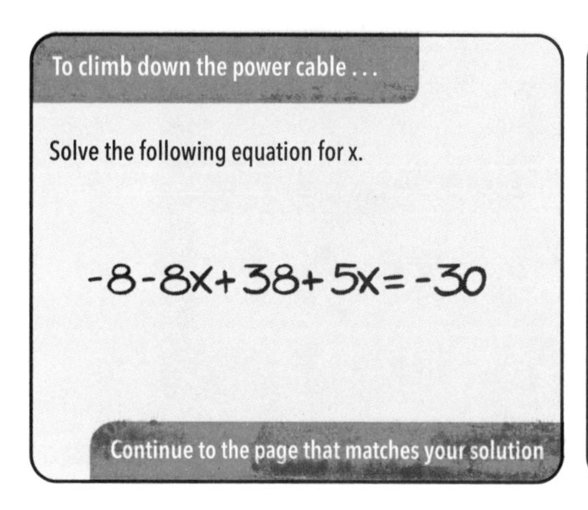

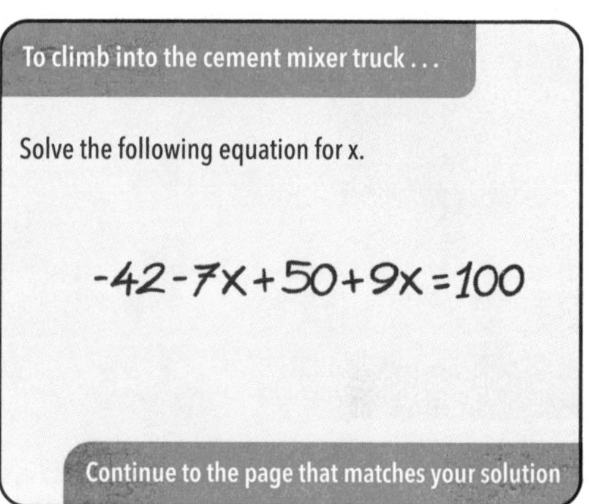

See page 55 of Adventurer's Advice Chapter 1 for help.

You should have arrived here from page 17

You swing your legs into the storm drain opening and quickly climb down the iron ladder into the cement tunnel below.

The storm drain is *just* tall enough for you to walk around, and it extends far to your left and to your right. The darkness in each direction is punctuated by dim light streaming through other storm grates above at street level.

To your right, you hear a loud SPLASH.

What was that?

You can continue down the storm drain toward the sound. That direction seems better lit. Or you can go the other way to avoid *whatever it was* that made that noise.

To head toward the splashing sound, continue to page 44.

To grope your way in the direction of darkness, continue to page 38.

You should have arrived here from page 6

Whoever this guy is, he hasn't been running through two hours of soccer practice every afternoon. You're in much better shape, and he's not keeping up. You see another guy in sunglasses join him, but they are both at least a block behind when you turn a corner, then duck into an alley.

Graffiti is scrawled on the brick wall in here. Red spray-painted words say, *Stop Dregg, Save America*. You have been seeing more and more protest graffiti about Dregg lately, but the lettering on this wall is impressive. You take a step closer, and as you do the *BEEP BEEP* from the wallet sounds. *Almost like an alarm*, you think. *Stop staring at a wall! Get GOING!*

Looking ahead, you see the alley is short—it dead-ends just ahead. Not good! You look around for options.

There's a storm drain a few feet away. You hurry over, hook your fingers into the little notch on the cover, and pull. It comes loose. Straining hard, you lift it just enough to see a metal ladder bolted to the wall, below street level. You could climb down that. But do you really want to? You have no idea what's down there.

You look around again. To your right there's a chain-link fence enclosing a construction site for a building that's still mostly steel framework. You could get lost in there . . . but can you climb over the fence before you get spotted?

You better choose fast.

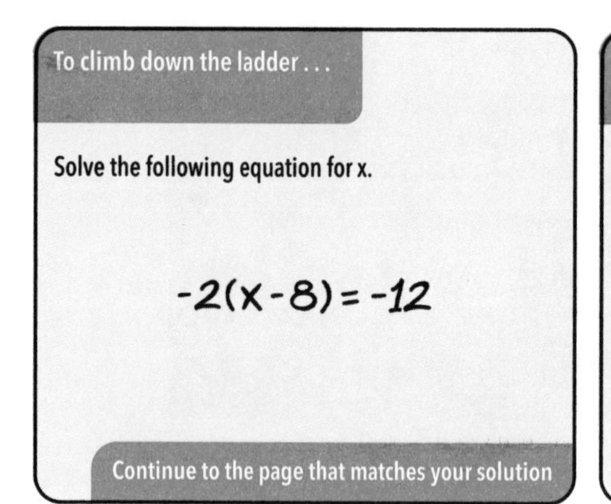

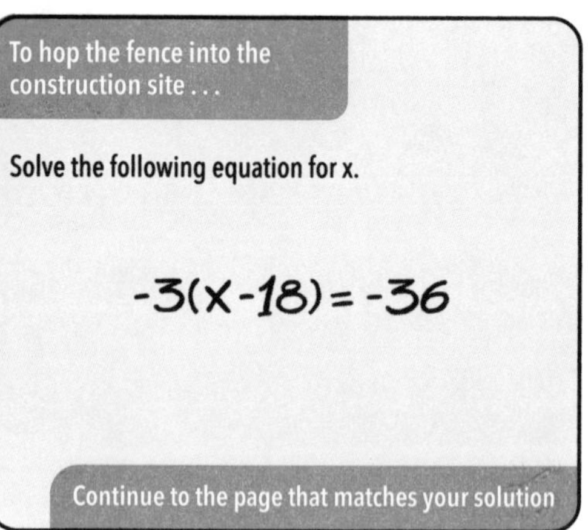

See page 54 of Adventurer's Advice Chapter 1 for help.

You should have arrived here from page 5 OR page 29 OR page 43

The tower is enormous. For some weird reason, the jingle from an old Dregg commercial pops into your head:

Dregg cupcakes are tasty and swell
And now they're free of chemical smells!

The cupcakes aren't too bad—but you have to wonder who's in charge of their advertising.

The glass exterior of the building is a shiny, sinister black. You push through the big revolving door into the lobby of Dregg Tower. You're met by the lowered arm of the automated entry gate. Fishing Doctor LaBella's ID card out of her wallet, you hold your breath as you slide it into the slot.

It works! A green light pops on, and the arm swings open.

You're in.

CHAPTER 1 GOAL: **Achieved!**

Good work!
This adventure will continue in Chapter 2!

You should have arrived here from page 14 OR page 35

You decide to head away from the splash. Carefully, you feel your way into the darkness. The walls begin to widen somewhat into a chamber that is maybe thirty feet across.

The walls are covered with enormous mushrooms and colorful lichen. The cement floor ends abruptly, and you find yourself looking down into an underground pond. It actually has fish! They're swimming your way in the water, as if they hope you've brought food.

They look kind of like sunfish, but bigger: sort of flat and oval-shaped. You spot a big one that might be over a foot long! It's hard to tell for sure. You lean forward. Maybe the low light and the water magnify their shape. But it's clear that these fish are hoping for food. Their eyes and mouths bob above the water.

Do they have little teeth?

To continue your adventure, turn to page 23.

Go back to the previous page and check your work. You should not have arrived here.

You should have arrived here from page 17

Scaling a chain-link fence is your kind of a challenge—you decide to climb it.

The fence surrounding the construction site is about ten feet tall. But you are in such good shape from soccer, you climb it easily.

High five for team sports.

You drop to the ground on the other side and scope out the building. It's about four stories. Steel girders outline where the walls will go. There's a rough wooden staircase. You jog that way, glancing around nervously. You haven't been spotted . . . yet.

You pass a row of smelly portable toilets and reach the stairs almost immediately.

They go up or down. Not too much light in either direction.

To take the stairs up, continue to page 24.

To take the stairs down, continue to the page 19.

You should have arrived here from page 22

YOW! The water is freezing. And deep.

As your plunge slows and you begin to rise back to the surface, you open your eyes and look around. There is an underwater culvert about fifteen feet away. You have no idea where it leads, *but neither do the goons on the bridge*.

It's almost dark when you break the surface. You gasp for air. The weight of your wet clothes drags you down a little.

There are streetlights on the bridge, and you hear yelling. Someone shouts, "Look harder!"

Right now you're hidden by the bridge—they haven't spotted you yet. But the river currents are pulling you downstream. You probably have a few seconds, no more, to figure out what to do before someone sees your bobbing head in the water.

Downstream on the right, not far off, you spot a small dock. The dock won't be as creepy as the underwater culvert, but it won't be as protected either.

You're a good swimmer; you could use the current and a strong front crawl stroke to get to that dock or the culvert. Hopefully you can get there before anyone can grab you. How far are these guys willing to go?

To swim for the dock, turn to page 17.

To swim to the culvert, turn to page 9.

You should have arrived here from page 45

You start walking away from the suits, trying to look casual.

"Hey kid," one yells, "that card is Dregg Corp property. Where'd you steal that wallet? And the card too. That doesn't belong to you. You're in big trouble, you know that?"

Steal the wallet? How do they even know about the card? Does it have a tracking chip? Your heart starts to race. Maybe you should take off and run for it. You're fast, and these two hulks in dark suits and leather shoes probably can't catch you. Probably.

Your phone vibrates with more messages. *Not now, everyone!* you think. You quickly scan the street. A tan Toyota is driving past. Should you wave them down? Or is it safer just to run?

If you run into the street to flag the car down . . .

Solve the following equation for x.

$$-6x-30=70-10x$$

Continue to the page that matches your solution

If you decide to make a run for it . . .

Solve the following equation for x.

$$-2x-22=26-5x$$

Continue to the page that matches your solution

See page 56 of Adventurer's Advice Chapter 1 for help.

You should have arrived here from page 7

Not far away, you spot the irregular shape of Dregg Tower rising up in the suburban landscape. Its gleaming walls of black glass are unmistakable. With a mixture of relief and anxiety—*okay, you didn't die, but are you really going to do this?*—you hurry that way.

You make it to Dregg Plaza. A tall statue rises from a fountain in the middle of a broad open square that's paved with flat stone. You've seen the statue before. It's a grandiose sculpture of Levi Dregg, Dregg Corp's legendary, mysterious founder.

Levi Dregg started the business thirty years ago, but he is never seen in public anymore. There are all sorts of stories about him; you have no idea what's actually true.

Beyond the statue is Dregg Tower. You take a deep breath and start that way.

See page 57 of Adventurer's Advice Chapter 1 for help.

You should have arrived here from page 14 OR page 35

You take a few shaky steps toward that splash.

Up ahead in the murky light, something moves.

It's hard to make out exactly what it is. But you *can* see that the sewer tunnel widens, and a big puddle has formed in the middle of this section of the tunnel.

You hug the right wall and step very carefully forward to avoid getting wet. You see movement again—and you freeze as you realize what it is.

In the center of this puddle, mostly submerged, maybe eight feet long . . . is an *alligator*. There's just no mistaking the dark, spiny ridges of its back, poking above the water.

Could someone have flushed a baby alligator down the toilet when it got too big to keep? Wasn't that an urban legend? How else did it get here? And what did it eat to get so big?

You swear it's got its eyes focused right on you. And they don't look friendly.

"I think the kid went into the storm drain!" you hear someone yell from above in the street. Time to move.

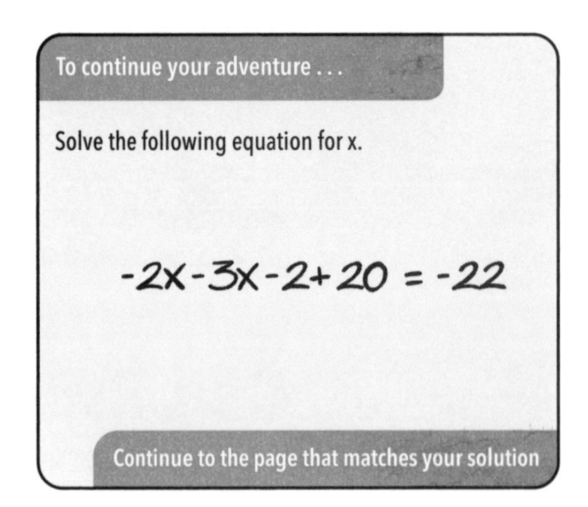

See page 55 of Adventurer's Advice Chapter 1 for help.

You should have arrived here from page 13

Taking this lost wallet straight to Dregg Tower is definitely your fastest way to the $1,000 reward.

Your phone buzzes with another text message. It's your mom.

Where are you? Is practice over? How about tacos for dinner?

I'm good, you type. But I might be a little late. Practice went long, and . . .

You pause, trying to think of what to tell your mom. But something strange is happening. It's almost . . . like you can feel an energy coming from the card in your hand.

That's not possible, right? You're imagining things. But maybe not. The "D" insignia on the card is glowing.

What is this thing?

You hear something and look up as two men in black suits turn the far corner and come your way. They are wearing sunglasses, even though the sun has nearly set. They look to be walking toward you, and one of them touches his ear with his hand. *Is he wearing an earpiece? Where did they even come from?*

You instinctively shove the wallet and card into your jacket pocket, and your phone too. You can finish and send the text as soon as you've figured out what's going on here. Because this is weird.

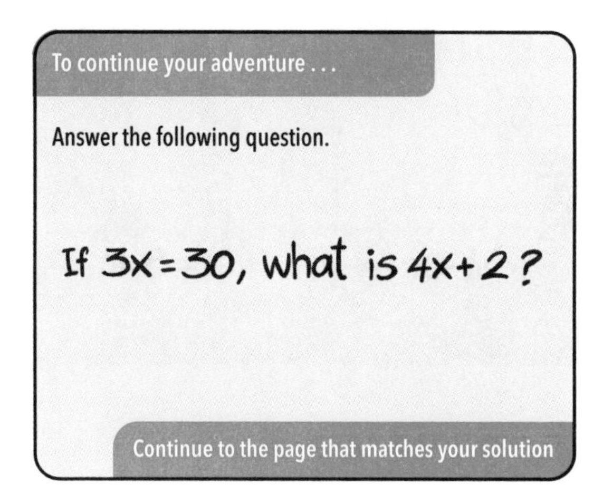

See page 51 of Adventurer's Advice Chapter 1 for help.

You should have arrived here from page 34

From the sound of squealing tires in the lower levels of the parking garage, you suspect the Dregg goons are getting closer. You climb up the long metal chute and into the cement mixer. The big mixing bowl has a wide mouth and deep walls, and you can't see inside. Unfortunately the only way in is head first.

The vans are coming, and you slide into the bowl to avoid being spotted. The mixer is filled with about a foot of wet cement, and you sink in up to your knees. You stay completely silent while the vans circle outside. The metal walls of the mixer seem to be blocking the signal coming from Donda LaBella's wallet, and after a few minutes of waiting, you can hear the vans take off back to the street below.

Unfortunately, the metal walls also block the outgoing signal from your phone, and the cement has hardened around your feet. You can't move.

It's going to be a long cold night in the cement mixer.

The End . . . for now.

You should have arrived here from page 4

They'll never think to check the ice cream truck. Plus you're friendly with Miss Amy, who has driven it for your entire childhood.

The truck almost seems to slow down for you to hop on! You are able to leap onto the back, hidden by the shadow of a tree.

Your phone vibrates. You glance at it. It's your mom.

Brownies for dessert! Where are you?

You barely feel the crack on the head that knocks you out. You do not have time to type that you won't be back for a long time. In fact, it will be days before you even come to. When you do, you are lying in a hospital bed, and finding Donda LaBella's wallet feels like a distant memory.

"It's real, Mom! I swear. It was a special Dregg ID card. It glowed. And had a tracking chip and a thousand-dollar reward."

"I'm sure it did, honey," she says, patting your hand.

The End

You should have arrived here from page 21

Your taxi breaks free of the traffic, speeding up the avenue. Looking back again, you see the van is still behind you. It's speeding too.

I don't get it, you think. *How could they know I've got this wallet? Does it have some tracking device? And if it does, WHY?*

"Uh-oh, kid—we got trouble," the driver says. "Look up ahead."

A couple of blocks up, another Dregg van has pulled across the avenue and stopped. It's blocking your lane.

"I'm not gonna be able to get through," the driver says to you, and you know that she's right.

There's a left turn coming up, and it leads onto a bridge across the river. Taking that turn would avoid the roadblock—and the bridge heads right toward Dregg Tower. Do you want to go that way?

Up ahead on the right is a parking garage. You could ask the driver to pull in there, then hop out to get lost among the cars.

"Tell me quick, kid! What are we doing?"

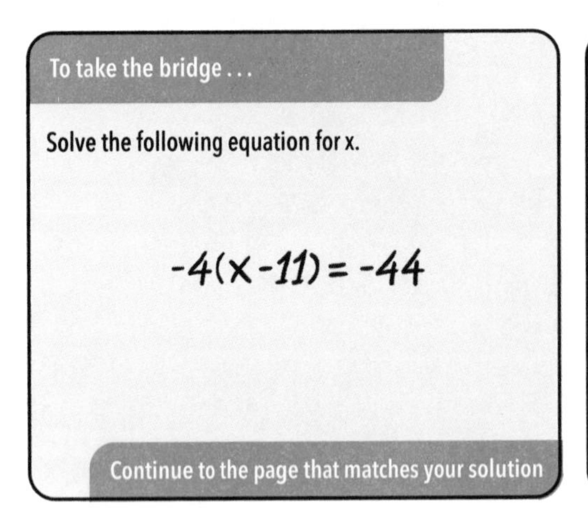

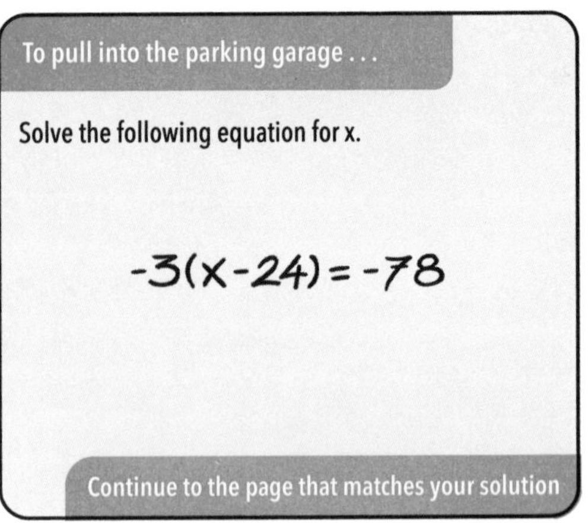

See page 54 of Adventurer's Advice Chapter 1 for help.

You should have arrived here from page 25

You scan the horizon for options. A couple more cars have come up behind you. This traffic is keeping the goons in the van at a distance. For now.

There's a bridge on your left that crosses the river, and Dregg Tower is just on the other side.

There's also a parking garage up ahead on the right. Maybe you could disappear inside? You could hop out and hide, while the two goons chase this lady all the way to her kids' ballet class.

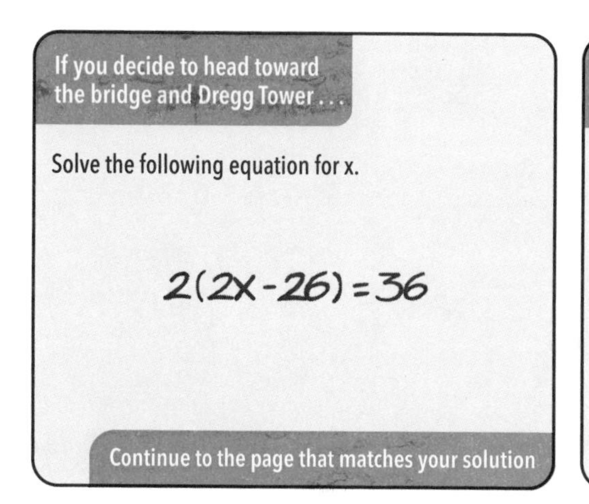

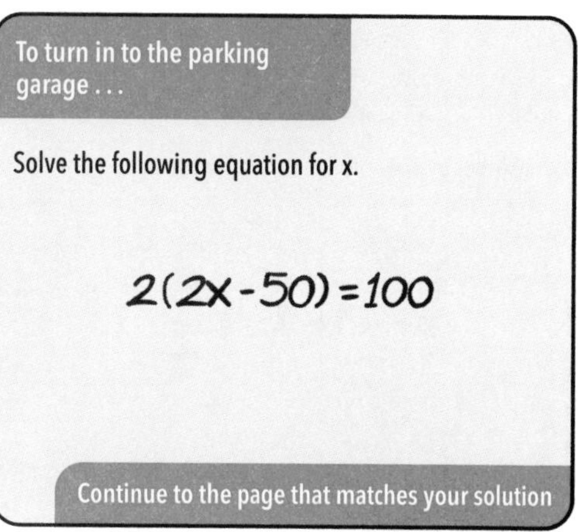

See page 54 of Adventurer's Advice Chapter 1 for help.

You should have arrived here from page 48 OR page 49

"Quick, can you take me into that parking garage?" you ask your driver, pointing toward its entrance.

"Got it," she nods. Driving expertly, she speeds ahead, then veers sharply right and jolts to a stop at the ticket-dispensing machine while rolling down the window to seamlessly yank out the little paper slip that shows what time you entered. *How often does this lady get into car chases?*

The gate lifts up and she shoots forward, past a sign that reads:

TOP LEVEL CLOSED

"What do you think we should do?" the driver asks you. "Those people saw us turn in here—they'll be behind us in a second."

You could ask her to drive up the ramps as high as you can go. Those goons in the van won't be sure which level you chose. But if they stay on your tail, you could get trapped up there.

You can also hop out here and hide among these parked cars. There are a lot of parked cars. It'd be hard to find you—but not impossible.

To hop out and hide among the cars . . .

Use order of operations to simplify the following expression.

$$5+(3+2)^2-\frac{8}{4}-2$$

Continue to the page that matches your solution

To continue up the parking garage . . .

Use order of operations to simplify the following expression.

$$8+(3+1)^2+\frac{10}{2}+5$$

Continue to the page that matches your solution

See page 52 of Adventurer's Advice Chapter 1 for help.

Using one-step equations (pages 28 and 45)

This problem type is asking us to figure out what number makes the first equation true (in this case, what number do we multiply by 3 to get 30). Once we have found that number, *then* we can plug that same number into the second expression.

If $3x = 30$, what is $4x + 2$?

$$3x = 30$$
$$\frac{3x}{\div 3} = \frac{30}{\div 3}$$
$$x = 10$$

STEP ONE: Solve the first equation. When a variable is up against a number (like 3x in this problem), they are being multiplied. We can use inverse operations to find the value of x. 3 is being multiplied by x, so we can divide both sides by 3. So x = 10, but we're not done!

STEP TWO: Substitute and solve. Because we know that x = 10, we can plug in 10 for x and solve. Order of operations tells us to multiply first (4 times 10), and then add 2 last.

$$4x + 2$$
$$4(10) + 2$$
$$40 + 2$$
$$42$$

The solution to this problem is 42, so continue to that page!

ANSWER KEY FOR THIS TYPE OF PROBLEM:

Page 28: Continue to page 18
Page 45: Continue to page 42

Order of operations (pages 14, 22, 30, and 50)

Order of operations! These problems are asking us to take a number sentence and make it simpler. Sometimes, though, our number sentence has more than one step, so where do we start? Remembering the acronym PEMDAS will help: Parentheses (P), Exponents (E), Multiplication and Division (MD), Addition and Subtraction (AS).

$2 + 3 \cdot (4 - 2) + 6^2$

$2 + 3 \cdot (4 - 2) + 6^2$
$\downarrow$
$2 + 3 \cdot 2 + 6^2$

STEP ONE: Parentheses. The first thing we look for is any math inside parentheses. In this problem, 4 – 2 is in parentheses, so we would combine those numbers (giving us 2).

STEP TWO: Exponents. Next we look for exponents. This problem has 6^2, and 6 to the power of 2 is 36. Let's change that before we move on to Step Three.

$2 + 3 \cdot 2 + 6^2$
$\downarrow$
$2 + 3 \cdot 2 + 36$

$2 + 3 \cdot 2 + 36$
$\downarrow$
$2 + 6 + 36$

STEP THREE: Multiplication and Division. If we have both, we just go from left to right. In this problem, we would multiply 3 by 2 (giving us 6). There is no more multiplication or division in this problem, so we are ready to move on to Step Four.

STEP FOUR: Addition and Subtraction. Again, if we have both, we go from left to right. In this problem, what we have left is the 2 + 6 + 36. Add those together to get 44, and continue to that page!

$2 + 6 + 36$
44

ANSWER KEY FOR THIS TYPE OF PROBLEM:

Page 14 (left): Continue to page 44
Page 14 (right): Continue to page 38
Page 22: Continue to page 41
Page 30 (left): Continue to page 24
Page 30 (right): Continue to page 19
Page 50 (left): Continue to page 26
Page 50 (right): Continue to page 34

Two-step equations (page 1)

This problem is a visual version of a two-step equation. We want to find out how many donuts come in each box. There is some number (x) of donuts in each box, and we have 3 boxes (3 x's). Knowing that, we can see this problem as an equation (3x + 4 = 43). This is an equation that we can solve!

$= 43$

$$\begin{array}{r} 3x + 4 = 43 \\ -4 \quad -4 \\ \hline 3x = 39 \end{array}$$

STEP ONE: Get rid of the 4 extra donuts. If we take away the 4 extra donuts, we will know how many donuts are in the boxes. To keep our equation equal on both sides of the equation, we need to subtract 4 on the right side too. This tells us that there are 39 donuts in the boxes!

STEP TWO: Find out how many donuts are in each box. If we divide the 3x by 3, it will isolate our variable, and it will tell us how many donuts are in each box! We also have to divide by 3 on the right side of the equation to keep both sides of our equation equal. There are 13 donuts in each box, so continue to that page!

$$\begin{array}{r} 3x = 39 \\ \div 3 \quad \div 3 \\ \hline x = 13 \end{array}$$

ANSWER KEY FOR THIS TYPE OF PROBLEM:
Page 1: Continue to page 13

Equations and distribution (pages 32, 36, 48, and 49)

This equation is asking us to figure out what number x has to be so that the left side of the equation equals 12. There are a couple ways we could solve this, but one good strategy is the distributive property.

$3(x-10)=12$

$3(x-10)=12$

$3x-30=12$

STEP ONE: Distribution. We want to multiply 3 by (x – 10). The distributive property lets us rewrite that number sentence by multiplying the 3 by each term inside the parentheses. So 3(x – 10) can be rewritten as 3x – 30.

STEP TWO: Get rid of 30 with inverse operations. From here, we will use inverse operations to isolate the variable. The 3 is part of the same term as the variable, so it will be easier to get rid of the 30 first. Right now, we are subtracting 30, and the inverse operation to subtraction is adding. If we *add* 30, it will cancel with the *subtract* 30 that we already have, and both 30s will go away. If we add 30 on the left, though, we need to do the same thing on the right so that the two sides of the equation stay equal.

$$\begin{array}{r} 3x-30=12 \\ +30 \quad +30 \\ \hline 3x = 42 \end{array}$$

$$\begin{array}{r} 3x=42 \\ \div 3 \quad \div 3 \\ \hline x=14 \end{array}$$

STEP THREE: Get rid of the 3 with inverse operations. We're close! The last number we need to get rid of is the 3. In this problem, the 3 is being multiplied by the variable, so to get rid of it we will need to divide. Dividing by 3 cancels with the "multiply by 3" that we have in the problem and leaves us with just the variable! Again, if we do something on one side of the equation, we have to do the same thing on the other side, so we have to divide our 42 by 3 as well. This gives us 14, and we will continue to that page.

AUTHOR'S NOTE: This problem can also be solved by dividing both sides by 3 first. Either way will give you the same solution!

ANSWER KEY FOR THIS TYPE OF PROBLEM:

Page 32 (left): Continue to page 14
Page 32 (right): Continue to page 30
Page 36 (left): Continue to page 14
Page 36 (right): Continue to page 30
Page 48 (left): Continue to page 22
Page 48 (right): Continue to page 50
Page 49 (left): Continue to page 22
Page 49 (right): Continue to page 50

Equations with like terms (pages 9, 19, 24, 26, 34, and 44)

What a mess! Look at all the stuff packed into this equation. We can solve this, but a good first step would be to make the equation less messy.

$$-4x + 11 + 2x + 16 = -35$$

$$-4x + 11 + 2x + 16 = -35$$
$$-2x + 27 = -35$$

STEP ONE: Combine like terms. There are four parts of the equation on the left side, and we call those parts *terms*. They're all being added together, and if we want we can group together the ones that go together. We know 11 + 16 = 27, and if we take away 4 x's from 2 x's we will get −2x. Rewriting this equation as −2x + 27 = −35 makes it a little less messy.

STEP TWO: Get rid of 27 with inverse operations. First, take away 27 on both sides. Remember, subtraction is the inverse of addition, so the 27s will cancel out on the left and leave us with −2x. We need to subtract on the right, too, to keep the equation equal, and we end up with an equation that says −2x = −62.

$$-2x + 27 = -35$$
$$-27 \quad -27$$
$$-2x = -62$$

$$-2x = -62$$
$$\div -2 \quad \div -2$$
$$x = 31$$

STEP THREE: Get rid of the −2 with inverse operations. Think of this 2 as "negative 2," not "minus 2." It's being multiplied by our variable, so to get rid of it, we need to divide. If we divide both sides by "negative 2," we find out that our variable has to equal 31 for this equation to be true! Continue to that page.

ANSWER KEY FOR THIS TYPE OF PROBLEM:

Page 9: Continue to page 31
Page 19: Continue to page 15
Page 24 (left): Continue to page 20
Page 24 (right): Continue to page 3
Page 26 (left): Continue to page 4
Page 26 (right): Continue to page 10
Page 34 (left): Continue to page 20
Page 34 (right): Continue to page 46
Page 44: Continue to page 8

Variables on both sides (pages 18 and 42)

We have a variable on both sides of our equation in this problem. The best way to solve this type of problem is to first get the variables together on one side of the equals sign.

$$2x + 15 = -3 + 5x$$

$$\begin{array}{r} 2x + 15 = -3 + 5x \\ -2x \qquad\qquad -2x \\ \hline 15 = -3 + 3x \end{array}$$

STEP ONE: Get all the variables on the same side of the equal sign. Subtracting either 2x or 5x from both sides will work, but I'm going to pick 2x so I don't need to deal with negative numbers quite as much. If I subtract 2x on both sides, the equation that I'm left with is $15 = -3 + 3x$, and all the variables are on the same side of the equal sign!

STEP TWO: Get rid of the "negative 3" with inverse operations. The negative 3 is a separate term, so it'll be easier to get rid of first. If I add 3 to both sides, it will cancel out, leaving me with $18 = 3x$. Remember, we need to add on both sides to keep our equation equal.

$$\begin{array}{r} 15 = -3 + 3x \\ +3 \quad +3 \qquad \\ \hline 18 = 3x \end{array}$$

$$\begin{array}{r} 18 = 3x \\ \div 3 \quad \div 3 \\ \hline 6 = x \end{array}$$

STEP THREE: Get rid of the coefficient with inverse operations. The 3 in front of the variable is being multiplied by x, so to get rid of it, we need to divide on both sides. What we end up with is $x = 6$, so I'm going to continue to that page.

ANSWER KEY FOR THIS TYPE OF PROBLEM:

Page 18 (left): Continue to page 6
Page 18 (right): Continue to page 21
Page 42 (left): Continue to page 25
Page 42 (right): Continue to page 16

Formulas (pages 5, 29, and 43)

Where did all the numbers go? Formulas are one way to show relationships between variables, and sometimes it is helpful to solve them for a different variable. This question is asking us to isolate the variable y, and we will know that we're done when our y variable is alone on one side of our equal sign. Lucky for us, the steps are the same. We just need to make sure we solve as if there are numbers in the problem.

$$ax + by = c$$

$$\begin{array}{rcl} ax + by &=& c \\ -ax & & -ax \\ \hline by &=& c - ax \end{array}$$

STEP ONE: Get rid of the ax term with inverse operations. We are trying to get y by itself, and the ax term can be easily removed with subtraction. On the left, that leaves us with just by, but on the right we get c – ax. Even though that looks weird, we can't actually simplify it any further, because we don't have any numbers! The equation we're left with says by = c – ax.

STEP TWO: Get rid of the b with inverse operations. We're close! The y variable is being multiplied by b, so to get it all by itself, we can divide both sides by b. What we're left with is y = (c – ax/b). This looks messy, but without any numbers we can't simplify it any further. This matches the second answer choice, so I'm going to continue to page 37.

$$\begin{array}{rcl} by &=& c - ax \\ \div b & & \div b \\ \hline y &=& \dfrac{c - ax}{b} \end{array}$$

ANSWER KEY FOR THIS TYPE OF PROBLEM:

Page 5: Continue to page 37
Page 29: Continue to page 37
Page 43: Continue to page 37

Basic proportions (page 13)

Fractions! This specific type of problem is called a "proportion" because we have two fractions with an equal sign in the middle. This type of problem can help us change size by scaling up or down. It can also help us convert between different units, so it's an important one to know how to solve!

$$\frac{x}{14} = \frac{10}{5}$$

$$\cdot 5 \cdot 14\, \frac{x}{14} = \frac{10}{5} \cdot 5 \cdot 14$$

$$x \cdot 5 = 10 \cdot 14$$

STEP ONE: Get rid of the fractions. This problem would be much easier to solve if we didn't have all these fractions all over the place. The fraction bar tells us that we are dividing, so we can cancel out "divided by 14" by multiplying by 14 on both sides. We can do the same thing with the 5. After canceling out the 14s on the left and the 5s on the right, what we're left with is 5 times x on the left and 14 times 10 on the right.

STEP TWO: Finish solving with inverse operations. 14 times 10 is 140. If we want to get the variable by itself, we need to get rid of the 5, which is being multiplied. To cancel that 5, we need to divide by 5 on both sides, which leaves us with x = 28. Continue to that page!

$$x \cdot 5 = 10 \cdot 14$$

$$5x = 140$$

$$\div 5 \quad \div 5$$

$$x = 28$$

AUTHOR'S NOTE: You may have learned how to solve this type of problem differently. Some teachers use "cross multiplication," "the butterfly method," or "the rule of three" to explain proportions. If one of those strategies makes more sense to you, solve it your own way!

ANSWER KEY FOR THIS TYPE OF PROBLEM:

Page 13 (left): Continue to page 45
Page 13 (right): Continue to page 28

Proportions with binomials (pages 3, 4, 8, 10, 15, 20, and 31)

Fractions! This specific type of problem is called a "proportion" because we have two fractions with an equal sign in the middle. This type of problem can help us change size by scaling up or down. It can also help convert between different units, so it's an important one to know how to solve!

$$\frac{x+3}{6} = \frac{4}{3}$$

$$\cdot \cancel{6} \cdot 3 \quad \frac{x+3}{\cancel{6}} = \frac{4}{\cancel{3}} \quad \cdot 6 \cdot \cancel{3}$$

$$3 \cdot (x+3) = 4 \cdot 6$$

STEP ONE: Get rid of the fractions. This problem would be much easier to solve if we didn't have all these fractions all over the place. The fraction bar tells us that we are dividing, so we can cancel out "divided by 6" by multiplying by 6 on both sides. We can do the same thing with the 3. After canceling out the 6s on the left and the 3s on the right, what we're left with is 3 times (x + 3) on the left and 6 times 4 on the right.

STEP TWO: Simplify the equation with the distributive property. On the right, we have 6 times 4. That's 24. The left side is a little trickier. We are multiplying 3 by (x + 3) which means we need to multiply 3 by the x *and* by the 3 in the parentheses. That side reduces to 3x + 9.

$$3(x+3) = 4 \cdot 6$$

$$3x + 9 = 24$$

$$3x + 9 = 24$$
$$-9 \quad -9$$
$$3x = 15$$
$$\div 3 \quad \div 3$$
$$x = 5$$

STEP THREE: Finish solving with inverse operations. If we want to get the variable by itself, we can get rid of the 9 first by subtracting 9 on both sides. This leaves us with 3x = 15. Now we need to get rid of the 3 that is being multiplied by the variable. To cancel that 3, we need to divide by 3 on both sides, which leaves us with x = 5. Continue to that page!

ANSWER KEY FOR THIS TYPE OF PROBLEM:

Page 3: Continue to page 5
Page 4 (left): Continue to page 47
Page 4 (right): Continue to page 11
Page 8 (left): Continue to page 2
Page 8 (right): Continue to page 12
Page 10: Continue to page 11
Page 15: Continue to page 29
Page 20 (left): Continue to page 5
Page 20 (right): Continue to page 11
Page 31: Continue to page 7

Similar shapes (pages 6, 16, 21, and 25)

"Similar shapes" is what you call shapes that are the same shape but not necessarily the same size. All the angles are the same, and the sides are *proportional*. The trick to this problem is recognizing that you can solve it with a proportion!

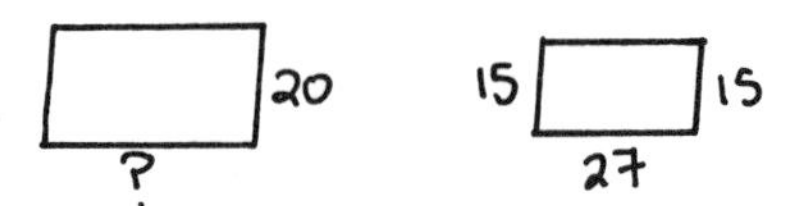

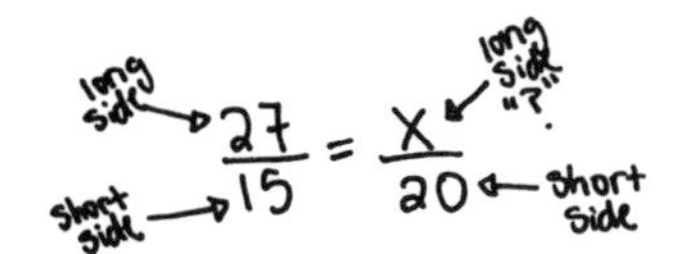

STEP ONE: Set up a proportion. The smaller shape on the left has a short side of 15 and a long side of 27. In other words, the long side is *almost twice* as long. If these shapes are similar, the other shape needs a long side that is *almost twice* as long as well! We can find the exact length by saying that 27/15 (almost 2) is equal to the missing side divided by 20. That relationship needs to be equal between the two shapes!

STEP TWO: Get rid of the fractions. This problem would be much easier to solve if we didn't have all these fractions all over the place. The fraction bar tells us that we are dividing, so we can cancel out "divided by 15" by multiplying by 15 on both sides. We can do the same thing with the 20. After canceling out the 15s on the left and the 20s on the right, what we're left with is 27 times 20 on the left and x times 15 on the right.

$$\cdot 15 \cdot 20 \; \frac{27}{15} = \frac{x}{20} \; \cdot 15 \cdot 20$$

$$27 \cdot 20 = x \cdot 15$$

$$27 \cdot 20 = x \cdot 15$$

$$540 = 15x$$

$$\div 15 \qquad \div 15$$

$$36 = x$$

STEP THREE: Finish solving with inverse operations. 27 times 20 is 540. If we want to get the variable by itself, we need to get rid of the 15, which is being multiplied. To cancel that 15, we need to divide by 15 on both sides, which leaves us with x = 36. Continue to that page!

ANSWER KEY FOR THIS TYPE OF PROBLEM:

Page 6: Continue to page 36
Page 16: Continue to page 32
Page 21: Continue to page 48
Page 25: Continue to page 49

CHAPTER 2

Dregg Tower's lobby is lit up brightly. It has gleaming white walls and a shiny floor with a polished, black-and-white checkerboard pattern. From a round white pedestal in the middle of the lobby, a hologram of Levi Dregg, the founder of Dregg Corp, waves at everyone who enters.

Well, that's creepy, you think. You're feeling a little uneasy after all you've been through so far, and this unsettling hologram doesn't help.

There's a floor-by-floor directory posted on the wall. Donda LaBella's card told you she is a Research Scientist—you see that Research and Development is on the eleventh floor. You wonder just what kind of research and development she's been doing.

Maybe you can find out.

CHAPTER 2 GOAL: **Reach the eleventh floor of Dregg Tower.**

At the elevator door, a line of people in Dregg polo shirts are waiting to have the badges on their lanyards checked by a tired-looking security guard. With one hand, the guard is inspecting the badges dangling from people's lanyards. With the other, he's holding a chicken burrito wrapped in wax paper, and it's dripping juice on the shiny floor.

Nervously, you join the line. If you flash Doctor LaBella's card quickly, maybe you can get by this guy. He has a lot going on. If you get caught, you could say Doctor LaBella is your mom, and she sent you in to grab something from her office.

Something like that, anyway. As the line grows shorter, the gears are turning in your mind.

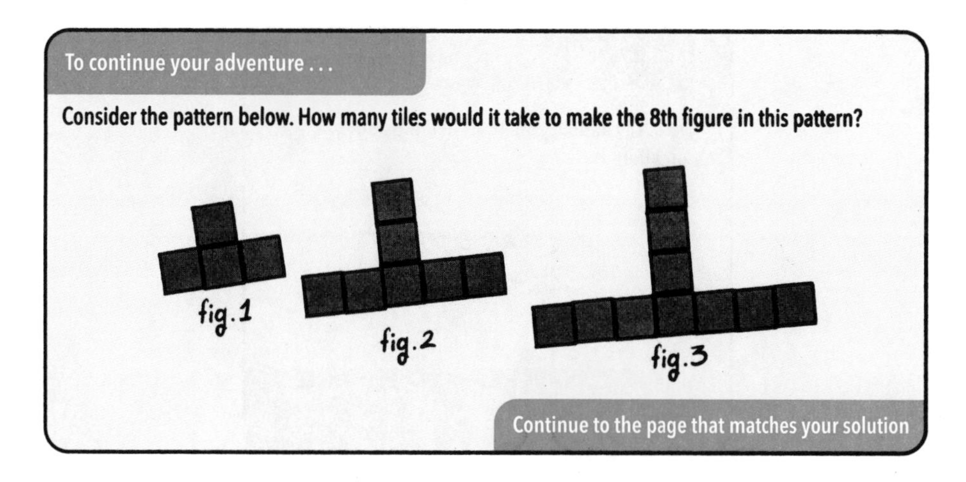

See page 51 of Adventurer's Advice Chapter 2 for help.

You should have arrived here from page 5

The shrink ray across the room continues to buzz—and you don't want to stick around to see what it can do. You turn and sprint toward the door you came in through.

You only make it two or three steps before you hear a loud *bzzzzzt*.

Some strange surge of electricity bursts through your body. Your muscles seize up for just a second, and you fall to the ground.

You jump back up and start running again for the door. But with each step, those swinging doors grow bigger and bigger.

When you finally get there, the doors tower above you. And even though you put your whole weight into pushing, you can't budge them. Not even an inch.

Something is really, really wrong here.

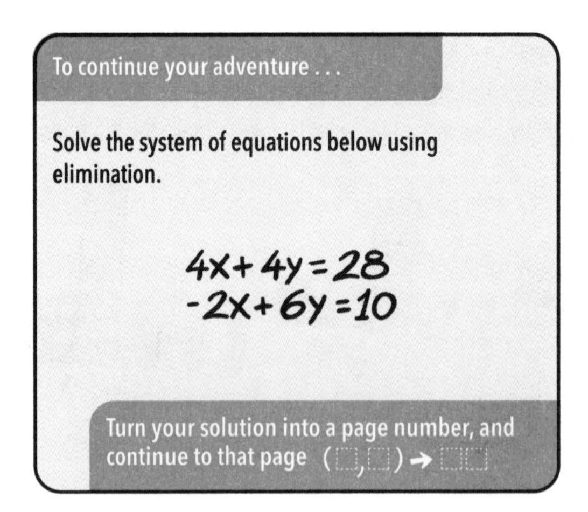

See page 58 of Adventurer's Advice Chapter 2 for help.

You should have arrived here from page 29 OR page 34

The elevator is still dark, but now that you have pulled open the elevator door, there is some low light illuminating the seventh floor below.

You crouch down and swing yourself out of the elevator onto the tile floor. It's only a five-foot drop, and you land quietly.

Quietly, but not quietly enough.

You hear a low growl and turn to see a shape that could be a dog. The light is dim. It *is* a dog, and a big one—a three-foot-tall Rottweiler.

The dog is wearing a metallic helmet, with colorful wires running from the helmet to a black backpack that is strapped around its torso. The Rottweiler curls its lips and bares its very sharp-looking teeth. But instead of a snarl, a tinny voice comes from the backpack:

"Hey, Food! Stay right there!"

By "food," the dog—or whatever is controlling its voice—clearly means *you*. Time to move.

There is an open door on your left. A sign on the door says:

MR. SIZZLEBOTTOM, REGIONAL MANAGER

You could make it inside and shut the door behind you. Or you could take off down the hall and hope to find the stairs. Either choice is better than ending up as dog food.

To head into the office . . .

Find the y-intercept of the line that connects these two points.

(3, 1) and (4, -3)

Continue to the page that matches your solution

To run down the hall . . .

Find the y-intercept of the line that connects these two points.

(2, 5) and (3, 3)

Continue to the page that matches your solution

See page 53 of Adventurer's Advice Chapter 2 for help.

You should not have arrived here

Go back to the previous page and check your work. You should not have arrived here.

CHAPTER 2

You should have arrived here from page 12

The Dregg corporation has a *shrink ray*? And are you the "lab rat"?

These are not the kind of questions you really want to be asking right now.

From across the room comes a buzzing sound. There's a large satellite dish like the one on the blueprint, and the sound it's emitting is growing. At the center of the dish is a glowing ball of twitching, pulsing electricity—and that's growing, too.

Uh-oh.

You'd better not stand there! You could duck behind the desk. Or you could make a run for the door.

To duck behind the desk . . .

Find the y-intercept of the line that connects these two points.

$(-5, -9)$ and $(2, 12)$

Continue to the page that matches your solution

To run for the door . . .

Find the y-intercept of the line that connects these two points.

$(5, 7)$ and $(-1, 1)$

Continue to the page that matches your solution

See page 53 of Adventurer's Advice Chapter 2 for help.

You duck quickly behind the desk, but its widely spaced legs give you no protection. From the shrink ray comes a loud sound, like "*bzzzt!*"

A sharp current of energy surges through your body.

But the blueprint said it was only a prototype—maybe it doesn't work. You peer out from under the desk—and to your total horror, the walls of the lab now tower above you.

So does everything else in here.

The ray worked, all right.

You make a run for the door, but in front of you an enormous shoe slaps down, blocking your path.

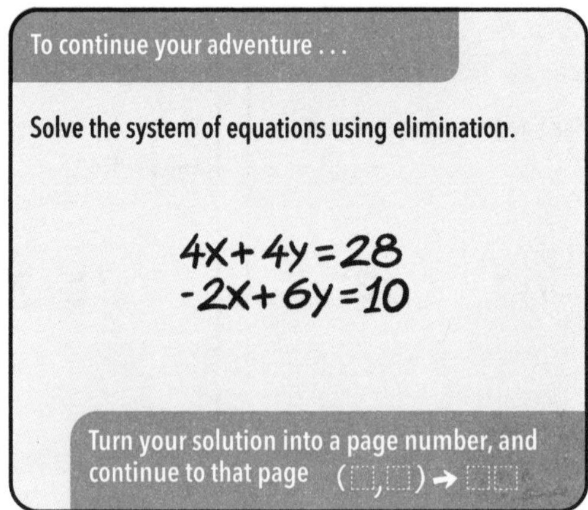

See page 58 of Adventurer's Advice Chapter 2 for help.

You should have arrived here from page 23

Your legs are *really* burning now, but you push on. After you've passed the fourth-floor landing and are halfway up to the fifth, a clattering from above stops you in your tracks.

At the landing above you is a robot. It has six long metal legs, each one ending in a sharp metal point. A long, hinged neck rises from its boxy body to a head with a video camera, watching your every move. The camera is topped by an unlit, pointy lightbulb.

"Scan complete," the robot says in a harsh, electronic voice. Its camera head swivels to point downward, toward you.

"Intruder detected!" The pointy light on the robot's head turns a bright red color and starts blinking. "Initiating termination!"

You leap back against the wall an instant before a laser beamed from the robot's body blasts a human-sized hole in the wall not two feet away. Cement debris from the blast sprays your shoes.

You better make a move, and *fast*.

You can retreat to the fourth-floor landing and push through that door.

Or you could climb through the hole the laser just made. Whatever is on the other side couldn't be more dangerous than this . . . right?

To climb through the hole in the wall . . .

Find the y-intercept of the line that connects these two points.

(-2, 15) and (3, 5)

Continue to the page that matches your solution

To head through the fourth-floor door . . .

Find the y-intercept of the line that connects these two points.

(-5, -2) and (-4, 0)

Continue to the page that matches your solution

See page 53 of Adventurer's Advice Chapter 2 for help.

You should have arrived here from page 7

The fourth-floor door takes you into a hallway. It's empty, and the silence is a little eerie. Hurrying down the hallway and away from the robot, you see an open door on your left. Hoping for the best, you duck inside.

The room is dark and seems just as empty as the hallway. Quietly, you close the door behind you.

You can just make out a long table of polished wood in the middle of this room. Fumbling on the wall by the door, you find a light switch and flick it on.

Six cushioned chairs are set around the table, and in each sits a robot just like the one that tried to kill you in the hall. They don't seem to notice you, at least not yet. Lights atop their heads are pulsing a soft yellow, and they're leaning across the table toward each other, as if they're communicating with one another. Did you just interrupt a private conversation?

How are you going to get out of here? On the wall beside you is a red fire alarm. *Maybe* you could pull that and slip away in the confusion. *Maybe.*

Or you could try to repeat the robot command you heard in the stairwell. It *might* confuse the robots enough for you to escape. It *might.*

Lots of "mights" and "maybes," but you are going to have to choose something soon. These robots are bound to notice you before long.

To pull the fire alarm . . .

Solve this system of equations using elimination.

$$14x - 3y = 27$$
$$-7x + 7y = 14$$

Turn your solution into a page number, and continue to that page (□,□) → □□

To try to repeat the robot's command . . .

Solve this system of equations using elimination.

$$-8x + 3y = -17$$
$$4x - 6y = -14$$

Turn your solution into a page number, and continue to that page (□,□) → □□

See page 58 of Adventurer's Advice Chapter 2 for help.

You should have arrived here from page 3

The dog is slobbering on the floor and talking through its backpack apparatus. "Don't move, I'm hungry!"

You sprint past and tear off down the hall. The dog is fast, but you have a head start.

You reach a pair of swinging double doors. You burst through them. As you do, the lights come back on around you.

Power's back on, you think. *Now where am I?*

You have entered some kind of laboratory. It's crowded with at least twenty large, tall glass tubes, and each tube is filled with glowing blue-green liquid. Each tube contains an animal, floating in its liquid and wearing a metal helmet and strapped-on black backpack.

There's a monkey. There are cats, racoons, sheep, a huge snake. One extra-large tube has a bison in it. A *bison!*

The dog will have the doors figured out soon, and you need a distraction. You spot a big bag of dog food propped up in one corner. *Maybe that will work.* There's also a heavy metal wrench lying nearby on a lab bench. You glance back and forth between the wrench and the glass tubes. Smashing one of the tubes would certainly count as a distraction . . . *Here comes the dog!*

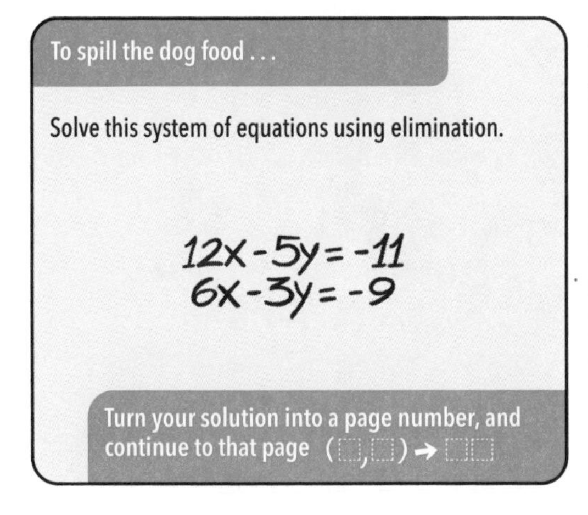

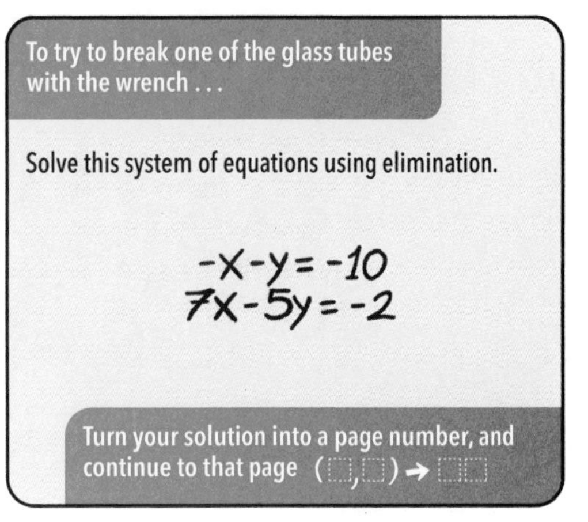

See page 58 of Adventurer's Advice Chapter 2 for help.

You should have arrived here from page 23

Enough with all these stairs. You pull open the door to the third floor, and you're immediately blinded by a bright light. An amplified voice is saying:

"*Passion. Commitment. Corporate acquisitions. The Dregg Corporation is making the changes that change how the world . . .*"

"*Cut!* Who let that kid in the shot?"

You seem to have stumbled into a commercial shoot. You are surrounded by cameras, and a big boom mic almost hits you in the face. Two actors in makeup are standing in front of a giant Dregg sign on the wall.

"Get that kid out of here!"

You scurry past the actors and the camera crew, down a long hallway. Before long, you spot the elevator.

You slip inside and hit the "11" button as the doors slide shut.

To continue your adventure, turn to page 29.

CHAPTER 2

You should have arrived here from page 7

You climb through the hole and there's not much light in here. As your eyes adjust, you realize you're in some sort of storage room.

And the walls on both sides are lined with robots like the one you saw on the stairs.

It's possible these robots are charging. The light on each camera head is blinking blue, and each robot body is connected with a thick black cable to a large steel box in the middle of the room. You look closer. A knob on the box is set to "low voltage: charge."

There is also a workbench against one wall. It's covered in robot parts. There are a couple of robot bodies on the floor beneath it. Each is actually larger than you, and each has a big open panel on its front. You can see they're empty inside, for now at least.

What to do? That robot is going to come searching for you any minute now. You could turn up the knob on the big box to "activate," hoping that might create a disturbance while you try to get away.

Or you could climb into one of those empty robot bodies to hide. You could easily fit inside.

Neither option feels very safe! But you've got to make a move *now*.

To turn up the voltage knob hoping this activates something . . .

Solve the following system by elimination.

$$-6x + 3y = 18$$
$$8x - 3y = -16$$

Turn your solution into a page number, and continue to that page (☐,☐) → ☐☐

To hide in the robot body . . .

Solve the following system by elimination.

$$6x - 7y = -24$$
$$-6x + y = -12$$

Turn your solution into a page number, and continue to that page (☐,☐) → ☐☐

See page 58 of Adventurer's Advice Chapter 2 for help.

You should have arrived here from page 29 OR page 34

The elevator is still dark, but now that you have pulled open the elevator door, there is some low light illuminating your face from the level above. You pull yourself up, out of the elevator, and onto the shiny tile of the eighth floor.

You stand and begin to walk. There is only one door in front of you, and you cautiously enter. The power surge knocked out the lights on this floor too, and it's hard to make out anything around you. You take a few steps—then the lights flicker and snap back on. There's only one door down this hallway, on the right. It isn't labeled.

You open the door and step into a giant round room. The nearest desk has a messy pile of papers on it. The paper on top looks like some sort of blueprint. The drawing on it shows a large satellite dish with a long metal cone sticking out from the middle of it.

Scribbled across the top of the blueprint are the words:

Dregg Shrink Ray

Prototype only

"All right, we've got power back," says a voice from across the room. "*And* it looks like we've got a new lab rat!"

To continue your adventure, turn to page 5.

You dash into the office and slam the door just before the Rottweiler can reach you.

"Why, 'ello there!" says a friendly voice from the shadows in the far corner.

As you squint to make out who is talking, the speaker takes a few steps toward you. It's a small pig, no more than eight inches tall. It's wearing a helmet and black backpack, just like the dog.

What is going *on* in this place?

You could try to ask this pig for help. *If it can talk, maybe it can answer*. Or you could try to escape another way.

Looking up, you spot a large air vent in the ceiling. Its grate looks easily big enough for you to get through. You could pull a chair over and try to climb in there.

What should you do—climb through the air vent, or talk to the pig?

To ask the pig for help . . .

Solve this system of equations using elimination.

$$-6x + 5y = -13$$
$$4x + 15y = 27$$

Turn your solution into a page number, and continue to that page (□,□) → □□

To climb into the air vent . . .

Solve this system of equations using elimination.

$$-15x + 6y = 27$$
$$-5x + y = 2$$

Turn your solution into a page number, and continue to that page (□,□) → □□

See page 58 of Adventurer's Advice Chapter 2 for help.

You should have arrived here from page 22 OR page 41 OR page 48

The service elevator is big and square inside, with a thick, wide pad attached to each wall. Dregg must use it to move bigger stuff, like equipment and machinery, up and down the tower.

There are long rips in some of the pads. Looking closer, you realize they're claw marks.

Huge claw marks.

The creature that made those must be enormous—and angry! A few of the elevator buttons are missing, but you jam the eleventh-floor button a couple of times until the doors slide shut. You'd better get this elevator moving before whatever did that comes back.

The elevator jolts to life, and this time there is no electrical disturbance to interrupt your ride to the eleventh floor.

You only have to wait a minute before the elevator makes a faint *ding* sound, and the doors slide open . . .

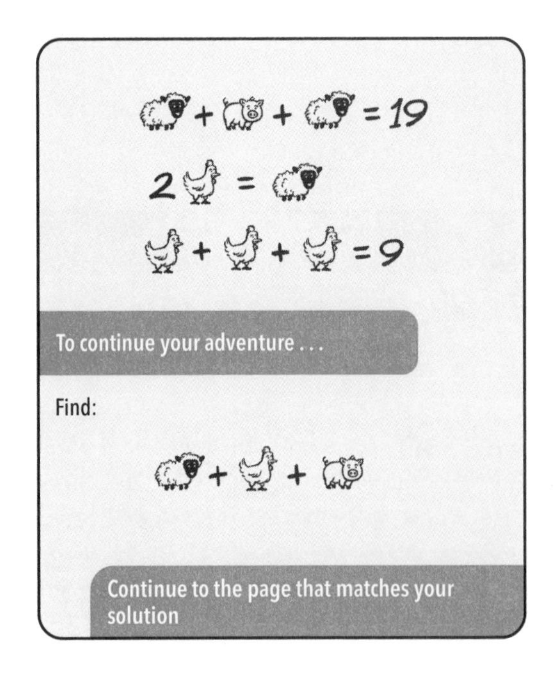

See page 55 of Adventurer's Advice Chapter 2 for help.

Go back to the previous page and check your work. You should not have arrived here.

CHAPTER 2

You should have arrived here from page 14 OR page 24

You step out of the service elevator on the eleventh floor.

Taking out Doctor LaBella's wallet and Dregg ID card, you walk across the tile floor, looking all around. Overhead, the fluorescent lights flicker a little.

You come to a pair of swinging double doors with a sign beside them:

DREGG RESEARCH AND DEVELOPMENT. AUTHORIZED PERSONNEL ONLY

Great work on Chapter 2!
This adventure will continue in Chapter 3!

You should have arrived here from page 13 OR page 31

The office has a desk, and you search its drawers until you find a screwdriver. Grabbing a chair, you unscrew the grate on the air vent. You climb up and wriggle through until you're in the ceiling. Just as you begin to move, you hear the whooshing sound of electronics turning back on—the power must have been restored.

Carefully you crawl over to another vent. You look down through the next vent and see you're directly above a scientific lab.

You peer downward. In the lab below you can see there are a number of tall glass tubes, and they are all filled with glowing blue liquid. One tube has a goat in it, floating amid the bubbles. It's wearing the same type of metal helmet and black backpack as the dog had.

There must be a dozen animals in those tubes. They all seem to be equipped the same way.

"Hey!" says a very small voice. "What are you doing up here? This is our turf!"

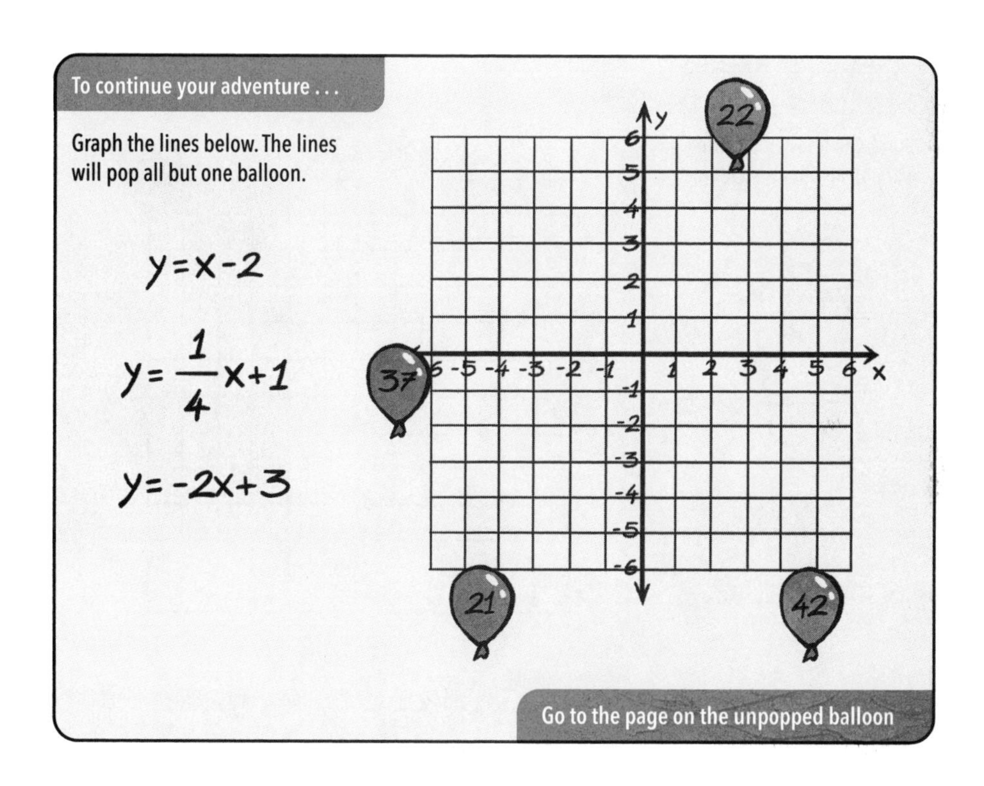

See page 54 of Adventurer's Advice Chapter 2 for help.

You should have arrived here from page 11

You grab the knob that is set to "low voltage: charge" and crank it hard to the right. You stop turning at "high voltage: danger."

Across the storage room, you see the robot that was chasing you. It followed you in here, and you duck to avoid being spotted.

The pointy lights atop the rows of charging robots are now flashing bright orange. The room is filled with at least fifteen orange lights, all flashing in unison.

The robots begin to chirp, "Er-ror. Er-ror." Dark smoke starts to billow from a few of the charging robot heads. The new voltage is too much for the robots to handle.

The smoke and the flashing lights are disorienting, but you still need to think fast if you want to get out of this room in one piece.

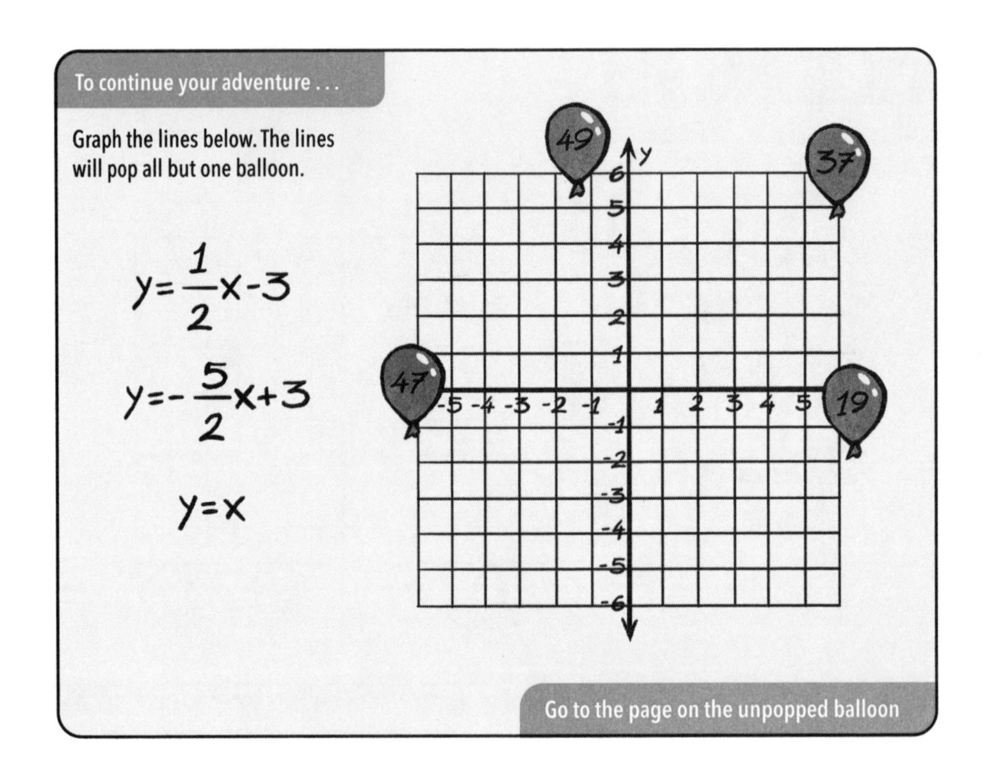

See page 54 of Adventurer's Advice Chapter 2 for help.

Go back to the previous page and check your work. You should not have arrived here.

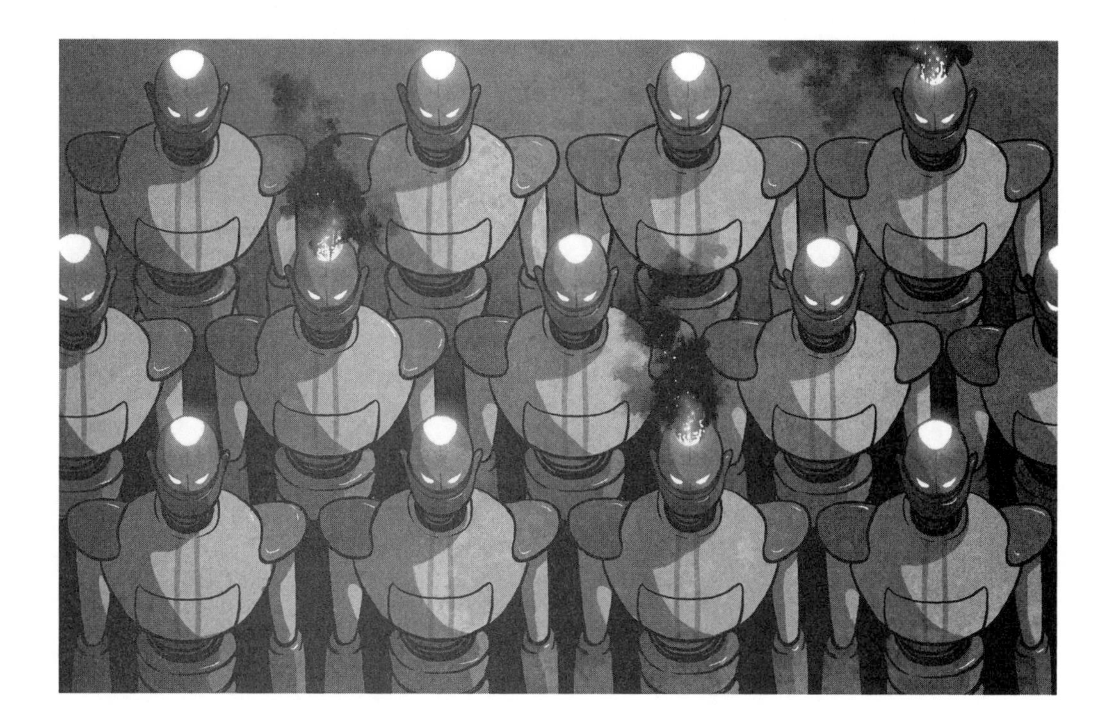

You should have arrived here from page 36

You *can't* sneeze—not now. You hold your breath, hoping the urge will pass.

But it doesn't.

"Ahhh-*CHOO!*"

"Intruder detected! Intruder detected! Search commenced." You hear the robot whirring into action.

Your hiding spot won't do any good against a laser, so you burst out into the storage room and run toward the door. The robot from the stairs cuts off your path. It has you cornered! You back up against the wall of the lab. You see no way to escape. You pull a beaker from the lab bench and throw it at the robot, but it just smashes into harmless little shards of glass. The robot doesn't even seem to notice.

Five thin metal arms fly out of the robot, each with knives at the end. They spin like fans.

"COMMAND RECEIVED, INTRUDER TO BE DECIMATED. REPEAT, INTRUDER TO BE DECIMATED."

The robot is inches away, and you have no way out.

The End

Go back to the previous page and check your work. You should not have arrived here.

You should have arrived here from page 17

"This is *our* turf," the voice repeats as two mice wearing tiny helmets scurry through the air duct toward you. One darts to the edge of the air vent you're climbing through and starts to chew vigorously.

You hear a loud pop, and the support for the air duct snaps, dropping the metal structure into the lab below. There's nothing to do but fall with it.

With a loud crash the metal tube lands on a lab bench, with you inside. Pain stabs through your knees and chest, but you crawl out and drop to the floor.

Painfully, you stand up and push through the doors to get out of this bizarre laboratory. You enter a different hallway than the one you came from. As you thought, the lights are all back on.

Down the hall is a service elevator. It's empty! Time to continue up the tower, before that dog finds you.

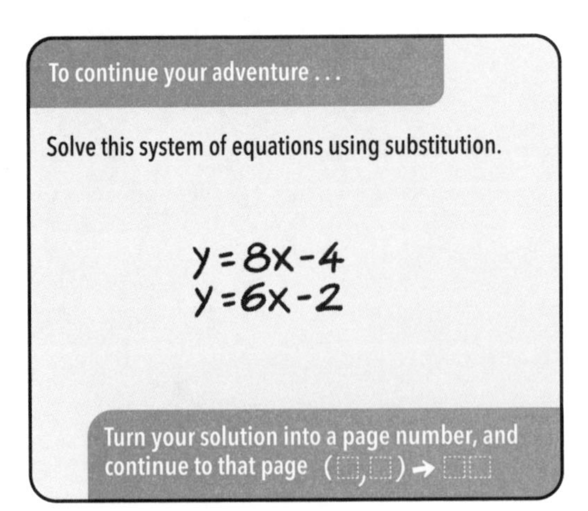

See page 57 of Adventurer's Advice Chapter 2 for help.

You should have arrived here from page 38

The stairs seem to go on forever. After climbing the second flight, you pass two Dregg workers coming down. You keep your eyes lowered, and to your relief they walk right by you. They seem to be too busy talking to notice that you're not wearing a company shirt.

"What is the deal with these power surges anyway?" one says.

"All I know is, they're happening more and more," says the other. "Some team in research must be using a *lot* of juice."

As you climb and they descend, the sound of their voices fades away. You're just coming up to the third-floor landing, and already your legs are starting to burn. Eight *more* floors of this? Maybe you should have taken the elevator after all.

Actually, you still could. You could leave the stairs at the third floor and go find the elevator. They're probably only checking badges in the lobby.

But if you leave the stairwell, people are bound to see you. The stairs above you look and sound empty.

To continue up the stairs . . .

Find the slope of the line that connects these two points.

(9, 13) and (5, -15)

Continue to the page that matches your solution

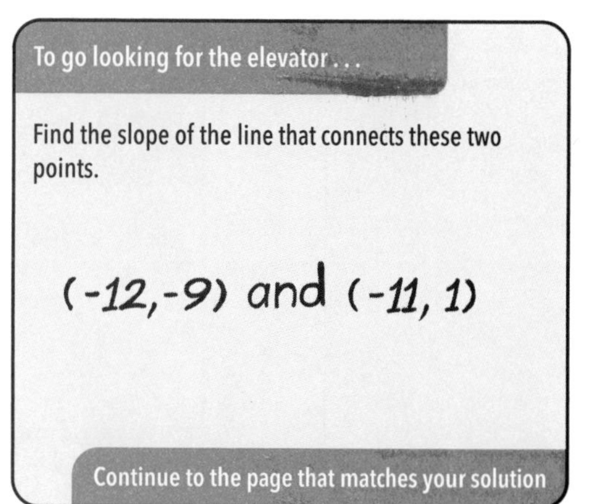

See page 52 of Adventurer's Advice Chapter 2 for help.

You should have arrived here from page 33, 44, 47, OR 50

As the doors to the service elevator slide shut, you jam your thumb against the button for the eleventh floor.

There is a mysterious purple mold on the button that sticks to your fingers, and as the elevator starts moving up, you notice that it's all over the floor too. The bottoms of your shoes feel sticky, and you quickly move to the back corner of the elevator. It's the only spot of floor that is free of the stuff.

As you continue your ascent to the eleventh floor, you notice some of the same purple fungus growing in a gigantic blob on the wall nearest you. It's slimy, and it's pulsating a little bit.

Maybe you would rather stand in the stuff, and you take a step back toward the center of the elevator car.

A soft ding signals your arrival, and you are ready to be away from this stuff. *It feels like the mold is watching you.* You glance quickly at the little lights over the door: eleventh floor.

The doors slide open . . .

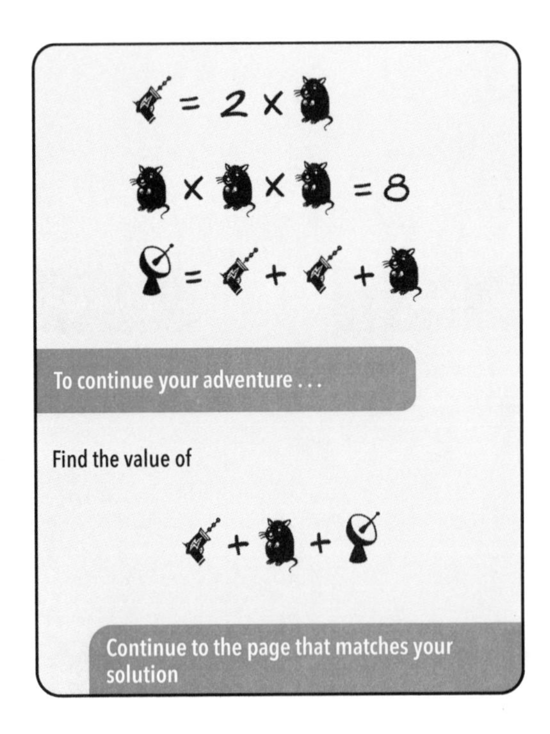

See page 55 of Adventurer's Advice Chapter 2 for help.

The security guard licks burrito sauce off his fingers. “A card? Sorry, eleventh floor is off limits,” he says. “If you don’t have a badge, you ain’t getting past here.”

Why don’t they let people onto that floor? And what are you going to do now?

The lights in the lobby flicker. “Not again,” the guard groans. “Third time today!”

The flickering grows intense. The checkerboard floor seems to shudder. *What is happening?* Suddenly the shuddering stops—and all the lights go out. The lobby of Dregg Tower is plunged into darkness. There is shouting all around you, and sounds of confusion.

The lights slowly return, and the guard is pacing, listening to a garbled message on his radio. Something about a *power surge*. For the moment, he seems to be distracted. This is your chance! The elevator doors are just starting to close. You could zip in there to join the crowd inside. You also spot a door to the stairs, over on the left and away from the guard. If you duck in there, you’re even less likely to be spotted. But do you really want to climb eleven floors?

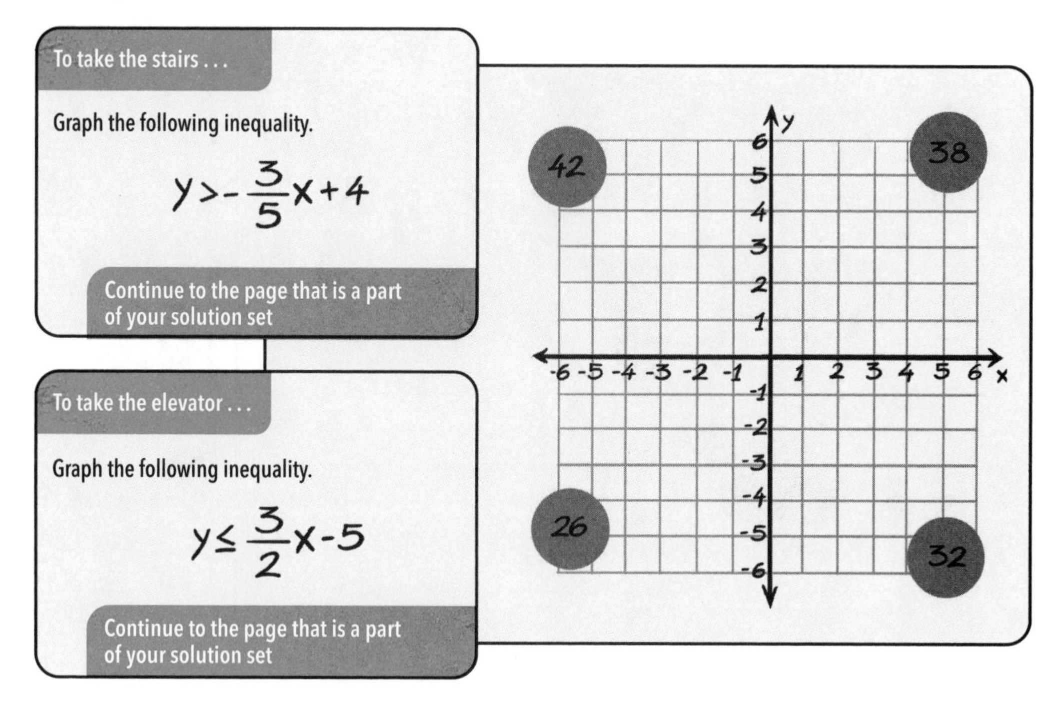

See page 59 of Adventurer’s Advice Chapter 2 for help.

Go back to the previous page and check your work. You should not have arrived here.

CHAPTER 2

You should have arrived here from page 9

The dog is through the doors now, but you have the dog food in hand. Hopefully he will eat this stuff instead of you! You rip the bag open, spilling kibble all over the floor.

The Rottweiler is running at you full speed and he slips on the little brown pellets of food. Flailing wildly, he smashes into a blue tube with a helmeted kangaroo inside. The collision causes the tube to rock back and forth alarmingly, and some of the blue goop sloshes out from the top.

The tube's momentum makes it rock harder—and as you watch in shock, the tall glass structure topples over. With a series of strange gulping sounds, a thick flood of blue goo glops out and slops onto the floor.

The kangaroo's tube smashes into the tube beside it, this one containing a leopard. This second tube crashes into the next tube, which crashes into the next one in a loud, goopy chain reaction.

To continue your adventure, turn to page 40.

You should have arrived here from page 43

The hamster wheel is loose on its frame, and it wobbles a little as you start to walk on it. Moving your legs is helping to calm your nerves. Soon you're running hard as the metal wheel spins around you.

"Now what?" you wonder aloud. "I need a plan . . ."

Your mind is cycling through your options, and you fail to notice that the wheel is wobbling badly. It starts rocking back and forth, and before you can stop your own momentum, the thing tilts over and crashes into one of the cage's glass walls. The glass shatters, spraying sharp shards all over the lab floor below.

You're free! You leap from the terrarium to the ground below. Unfortunately for you, the sound of the breaking glass alerted a giant guard dog wearing a strange backpack. He runs into the room, barking and drooling. At your size, the animal towers above you.

There is nowhere to hide, and your tiny little legs can't outrun this gigantic animal. You cover your face with your hands, but you're pretty sure of what's coming next. At six inches tall, you're the perfect doggie toy!

The End

You should have arrived here from page 10

Little lights above the elevator door show your progress: 4, 5, 6, 7 . . .

Suddenly the lights in the elevator start to flicker. The floor shakes under your feet. The elevator stops abruptly, and the lights go out. Even the little overhead lights that were counting floors are suddenly dark. The elevator is pitch black. *It must be another power surge.*

Minutes pass but the lights don't come back on, and the elevator doesn't budge.

You feel around with your hands until you are able to squeeze your fingers between the elevator's double doors, and they pull open with surprising ease.

During the power surge, the elevator stopped between floors. You could climb up to the eighth floor, or if you crouch, you could hop down to the seventh floor.

To hop down to the seventh floor, turn to page 3.

To climb up to the eighth floor, turn to page 12.

You should have arrived here from page 31

You grab the door handle and push it open. "Dinner!" the dog shouts through its backpack speaker. He lunges at you—but the little pig reacts quickly.

"*Bosco!*" the pig says. "We do not *eat* our guests!"

The dog's head dips, as if in shame. But still he emits a low growl: "But . . . huuuuungry."

"That's it!" hollers the pig, and it leaps toward the Rottweiler. The dog must outweigh the pig by 70 pounds, but it's still clearly scared. The dog leaps to avoid the pig that is now nipping at his hind legs.

The pig chases the dog down the hall and into a lab. More curious than scared now, you follow them inside.

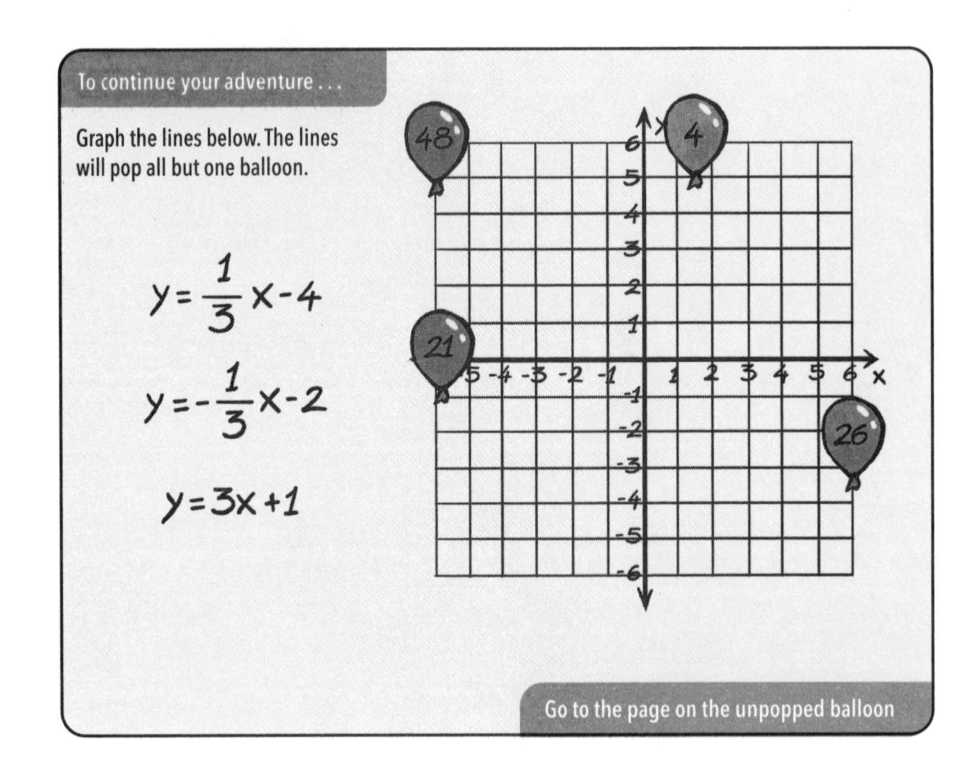

See page 54 of Adventurer's Advice Chapter 2 for help.

You should have arrived here from page 13

The little pink pig speaks again before you can say anything.

"That Bosco is such a bother. Always harassing our visitors," the pig seems to say, though its mouth, or snout, has not moved. "Open that door," it says, "and I'll have a word with him."

Peering closely while it talks, you realize the voice is coming from its little backpack. Just like Bosco. Does it contain some kind of voice generator? Is it voicing what the pig actually thinks, as transmitted from its helmet? Or is some other entity using the pig to speak?

And how weird *is* this place? You remember the sign on the door. Is this little pig the regional manager?

You're not sure if you should do what the pig says. Maybe you should keep the door shut and the dog out. You look back to the large air vent above you.

You could climb out through the air vent and try to reach another part of the tower. Or you can open the door, as the pig said, to let in the big dog waiting outside.

To climb into the air vent, turn to page 17.

To open the door, turn to page 30.

You should have arrived here from page 25

You slip into the elevator just as the doors are closing. The Dregg employees in here size you up briefly. You must look a little disheveled after your eventful journey to the tower! In suspense, you hold your breath, waiting to be asked a question.

But they just go back to their conversation.

"It's definitely happening more often," one employee says. They must be talking about that power surge.

"It's research," says another. As the elevator rises, people around you nod.

"Right," someone says in a strange tone. "*Research.*"

The elevator stops at the fourth and sixth floors, and everyone but you gets out.

Alone now, you punch the button for the eleventh floor. The door closes, and you're rising fast.

Pulling out your phone, you send Hazar a quick text.

I need you to find out anything you can about a Doctor Donda LaBella.

OK. Who she? Hazar replies.

Scientist, you tap out. At Dregg. Gotta run! Send what you find.

Hazar writes, You OK? But you don't have time to answer.

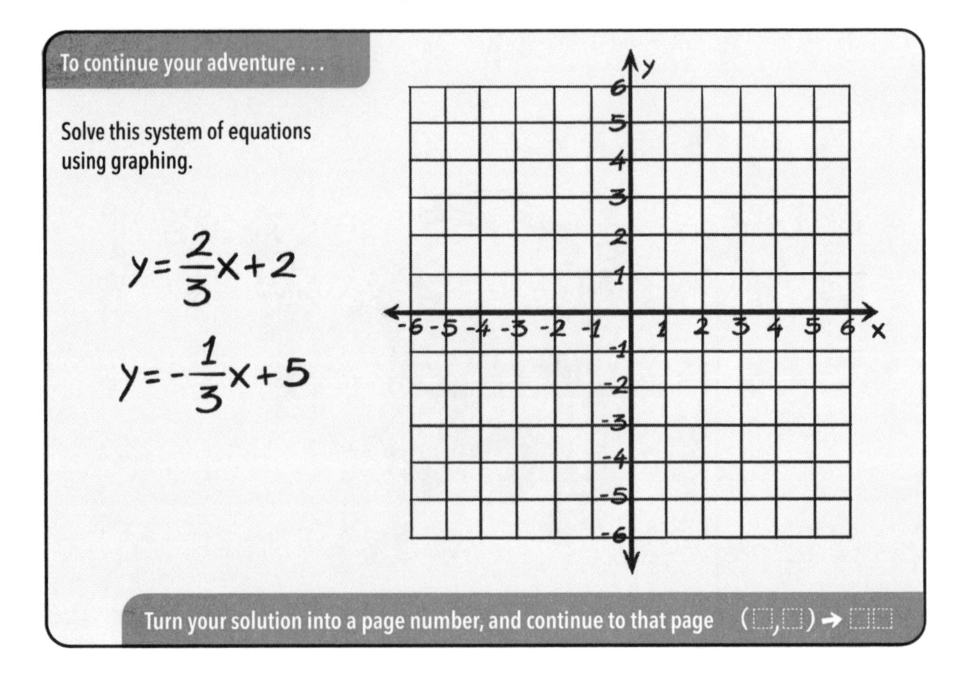

See page 56 of Adventurer's Advice Chapter 2 for help.

After the first laser blast, the other robots begin to turn against each other. "Intruder alert, intruder alert!" You have accidentally set in motion a laser fight, and you might not want to stick around to see how it ends. If those lasers can burn through metal and plastic, your jacket probably won't do very much to keep you safe.

You hurry out of the room—and as you shut the door behind you, more robot voices activate inside, and there's another explosion. *Those robots are on a roll.*

Rushing away down the hall, you come to the open door of a service elevator. Nobody is inside.

Whatever is up there, you think, *it can't be any weirder than what's happening here.*

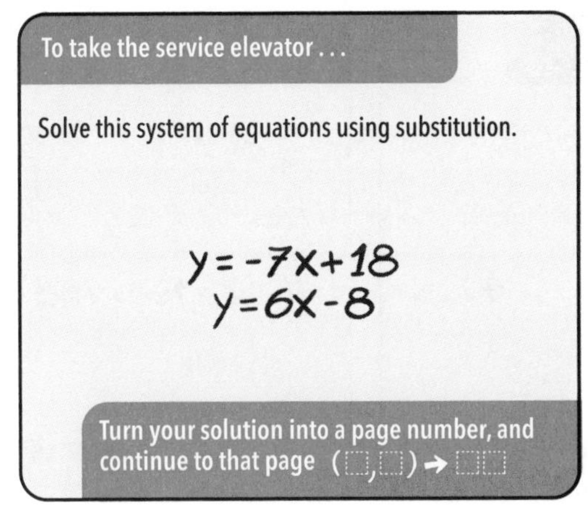

See page 57 of Adventurer's Advice Chapter 2 for help.

You should have arrived here from page 32

Little lights above the elevator door show your progress: 4, 5, 6, 7 . . .

Suddenly the lights in the elevator start to flicker. The floor shakes under your feet. The elevator stops abruptly, and the lights go out. Even the little overhead lights that were counting floors are suddenly dark. The elevator is pitch black. *It must be another power surge.*

Minutes pass but the lights don't come back on, and the elevator doesn't budge.

You feel around with your hands until you are able to squeeze your fingers between the elevator's double doors, and they pull open with surprising ease.

During the power surge, the elevator stopped between floors. You could climb up to the eighth floor, or if you crouch, you could hop down to the seventh floor.

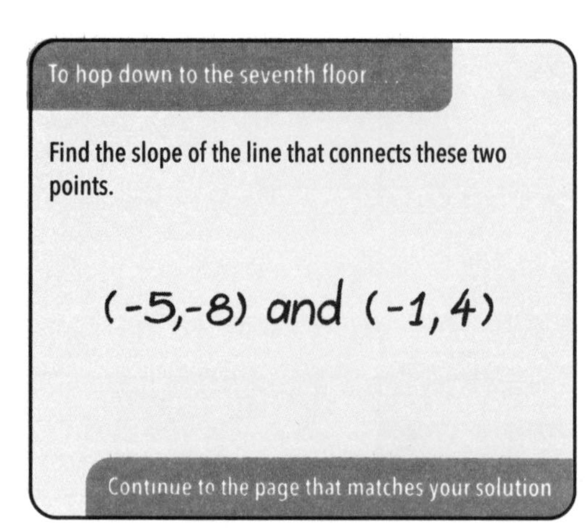

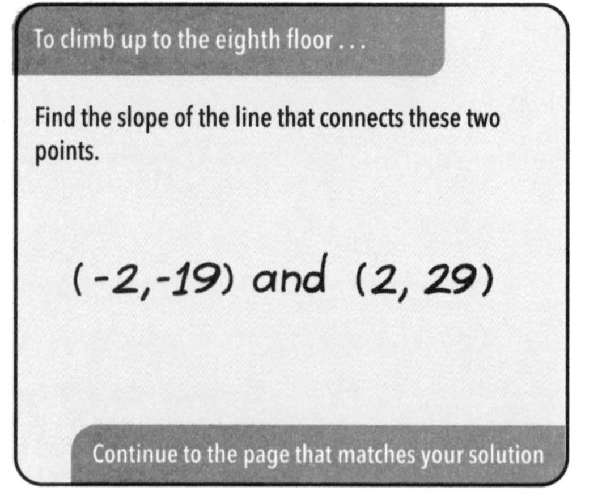

See page 52 of Adventurer's Advice Chapter 2 for help.

You should have arrived here from page 8

You reach for the fire alarm and pull down on its small white handle. The robots jerk upright, and they all rise from their chairs as the siren starts to blare. Should you hide? Should you run?

Their camera heads swiveling and lights blinking orange, one by one the robots spot you. They start to move your way just as water begins spraying from the sprinklers overhead.

Apparently the feet on the robots' spider-like legs were not designed for a wet floor. One loses its footing and topples over, smashing its camera. As water from the sprinklers continues to soak everything in sight, the other robots begin to twitch and glitch.

They aren't waterproof!

Smoke starts to rise from the robots' bodies. It looks like they're short-circuiting from the torrent of water!

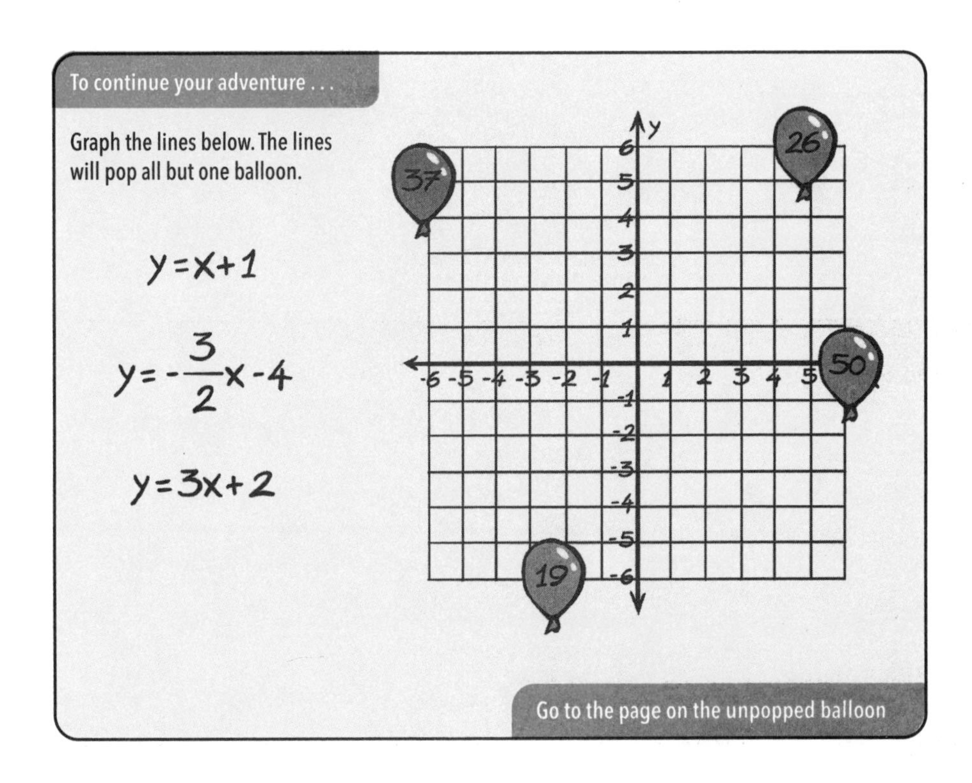

See page 54 of Adventurer's Advice Chapter 2 for help.

You should have arrived here from page 11

You climb into the empty robot body through its open front panel. Pulling the metal door closed from the inside, you huddle in a tight crouch, trying not to breathe and hugging your knees.

There's a whirring sound coming from the storage room beyond your metal hideout. The robot from the stairs must have entered the room.

"Scan-ning. Scan-ning," it chants in its odd, robotic cadence. You hear a different sound, like a small motor activating a mechanism. You guess that's the video-camera head, turning on its swivel to search the room.

You hold your breath. And even though you know it won't help, you also shut your eyes.

Oh no. You feel a sneeze coming on.

To continue your adventure, turn to page 20.

Go back to the previous page and check your work. You should not have arrived here.

You should have arrived here from page 25

You dash past the elevators, shove open the door to the stairwell, and close it behind you. In here, the silence seems to echo, and bare cement stairs zig and zag upward as far as you can see.

You've gone a few steps up the first flight when the lights flicker again, but this time they stay lit. As you climb, your footsteps are the only sound you hear.

Pulling out your phone, you send Hazar a quick text.

I need you to find out anything you can about a Doctor Donda LaBella.

OK. Who she? Hazar replies.

Scientist, you tap out. At Dregg. Got to run! Send what you find.

Hazar writes, You OK? But you don't have time to answer.

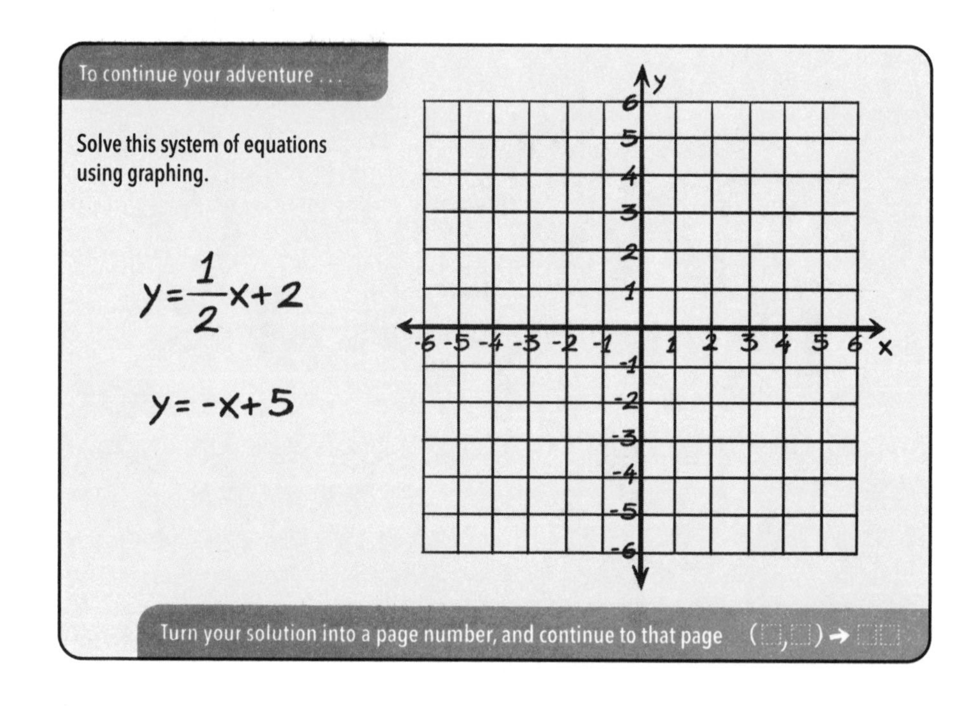

See page 56 of Adventurer's Advice Chapter 2 for help.

You should have arrived here from page 43

The water bottle towers over you. You try to take a quick drink, but so much water sloshes over your tiny body that your clothes are soaked. You stand back, trying to think up a plan.

The water bottle has ridges along its sides. That gives you an idea.

If you lean back against the ridges of the water bottle and push your feet up against the glass wall, you can inch your way upward. It's hard climbing, but by heaving and pushing yourself up inch by inch, you finally reach the top.

You grab the metal rim of the terrarium, pull yourself up, and look around. A light for the cage is clipped to the rim, with a power cord snaking down. Holding onto the rim very carefully, you work your way across until you reach the cord.

Grabbing it, you take a deep breath and start sliding down toward the floor of the room.

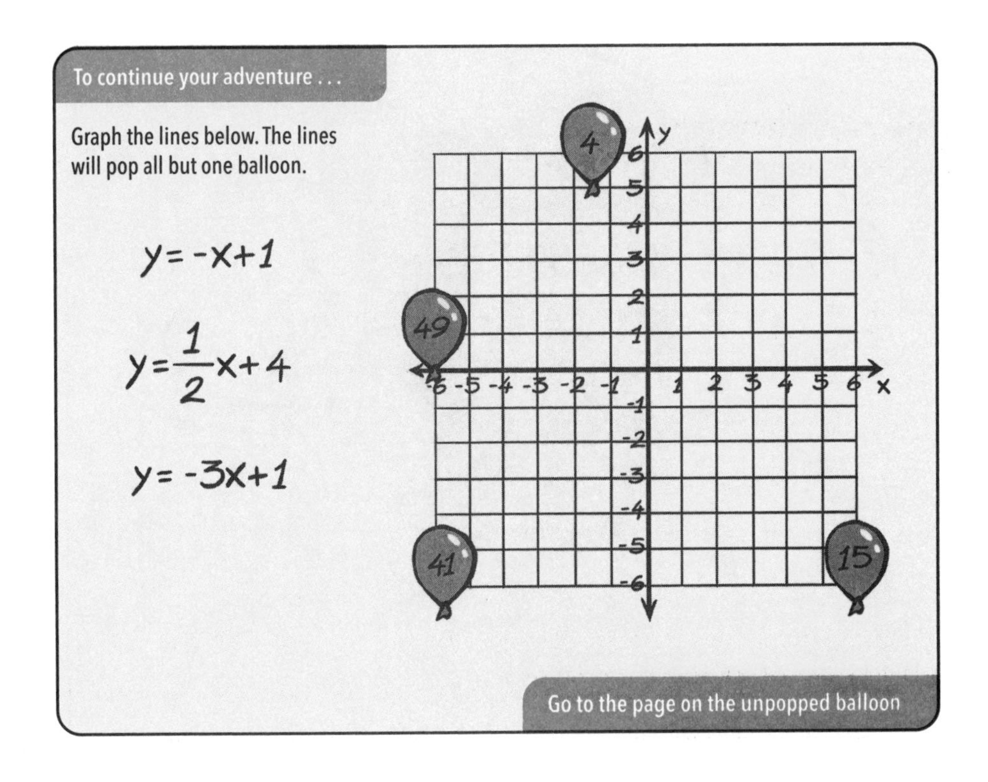

See page 54 of Adventurer's Advice Chapter 2 for help.

You should have arrived here from page 27

As each glass tube topples to the ground, it vomits out its weird fluid onto the floor. The animals too spill out across the floor of the lab, covered in goo.

One by one, the animals stagger to their feet. Their legs are shaky and they seem to be disoriented as they wake up from their time in the tubes.

You rush to the door against the back wall, but your hands are covered in the slimy blue ooze. You can't get a good enough grip to turn the handle!

A leopard shakes its head, sending little bits of goo flying across the lab. It's had a long sleep, and it woke up hungry. It fixes its gaze on the easiest meal in the room.

That would be you . . .

The End

You should have arrived here from page 39

You land on the floor kind of hard, but you're all right. The shrink ray is not far off—but you're so tiny, it takes you a full minute to reach it.

You use the power cable to climb up to the top of the satellite dish, where the controls are. One push-button is lit up, and it says: *Shrink*.

The button beside it says: *Unshrink*.

With all your weight, you jump on that push-button. It clicks down!

The shrink ray begins to charge another blast. You wrap your arms and legs around the power cord. It's a normal-size cord, but with your tiny size, it makes a perfect fire pole. Your tiny feet reach the ground, and you rush over to where the ray is pointing.

Just in time too. The blast sends a powerful energy current through you, just like the last time. You close your eyes and grit your teeth. Your body is . . . changing.

When you open your eyes, you're back to normal size.

Those two scientists haven't come back. You rush out the swinging doors and down the hall.

You come to a service elevator with open doors and no one inside.

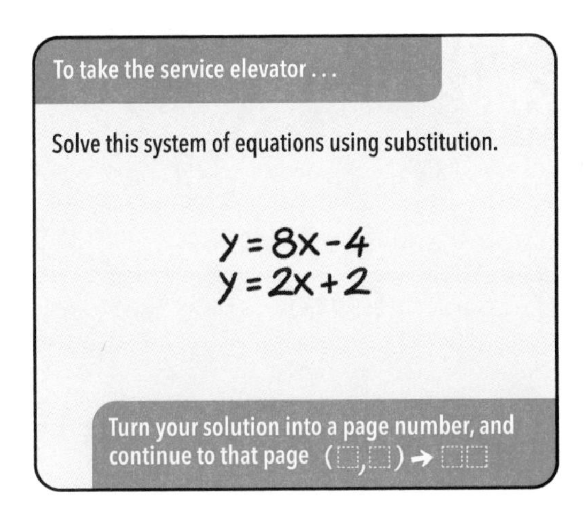

See page 57 of Adventurer's Advice Chapter 2 for help.

Go back to the previous page and check your work. You should not have arrived here.

CHAPTER 2

You should have arrived here from page 2 OR page 6

A giant hand grabs you and lifts you off the ground. You're at eye level with a towering man in a white lab coat. Your body is only as tall as his face. He doesn't say anything to you.

"Lab coat" carries you for a few steps, then places you into a glass terrarium. It's just like one Taylor has for her hamster: there's a water bottle upside down on one side, with a sipping tube, and in the middle of the floor stands a wire wheel. For your exercise.

Your feet sink into soft wood chips on the floor. You look out at what are now two giant faces, peering in at you. They must be scientists—they're scribbling notes on clipboards. After a few minutes they turn away and walk out through the main door into the hall.

NOW what? You could try to scale one of the glass walls, but you're way too small and they are way too smooth. It might be possible, though, to climb up between the water bottle and the wall it's clipped to.

Getting shrunk really shook you up! You're feeling jittery, and your thoughts are almost racing out of control. If you could just settle down a little, maybe you could come up with a better plan.

What if you hopped on the hamster wheel and ran for a few minutes? That might calm your jitters and settle your mind enough to think straight.

To try to climb the water bottle, turn to page 39.

To go for a run on the hamster wheel, turn to page 28.

You should have arrived here from page 46

Blue goop is dripping from the sheep's thick woolly coat. "Bosco, you sellout! You have been outside of your tube for weeks, and this little human freed me in a moment!"

"I don't have thumbs!" answers the dog. "What was I supposed to do?"

But the animals' mouths are not moving. Their voices are coming from the packs strapped to their backs. You can see that each one has a small speaker.

"Well, if you were a good boy, you would have figured *something* out . . ."

As the two animals argue, you slowly back away and into the hall.

You start to run.

You come to a service elevator. Its door is open, and you step inside.

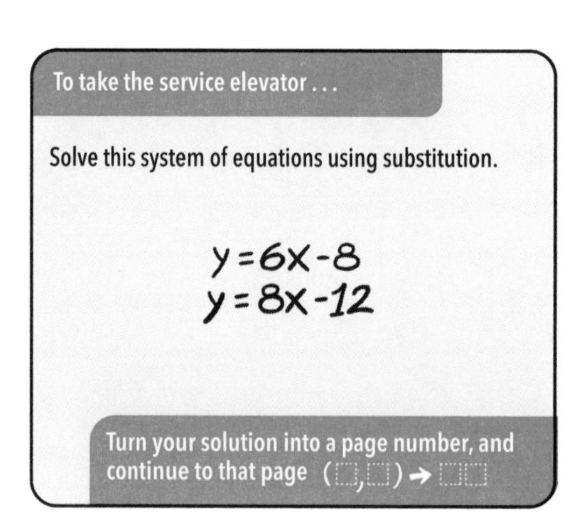

See page 57 of Adventurer's Advice Chapter 2 for help.

You should have arrived here from page 8

The robots turn in their chairs to face you. Remembering how the one in the stairway sounded, in an edgy metallic voice you say, "Scan complete. Room secure."

The pointy light on one robot's head glows yellow. "Intruder alert," it says.

"Room secure," you repeat.

"Intruder alert."

"Room secure." Why are you wasting your time arguing with this robot? It has a laser!

Right on cue the robot's light turns red.

"Intruder detected! Eliminate!"

You're about to be vaporized for sure. But rather than locking on you, the robot's camera head is still swiveling, still searching. It stops, its red light shines brighter—and it shoots a laser beam across the table. Direct hit! Another robot's head explodes, sending shards of metal and plastic spattering against the other robots.

"Termination complete," the first robot declares.

But now the lights of all the *other* robots are turning red.

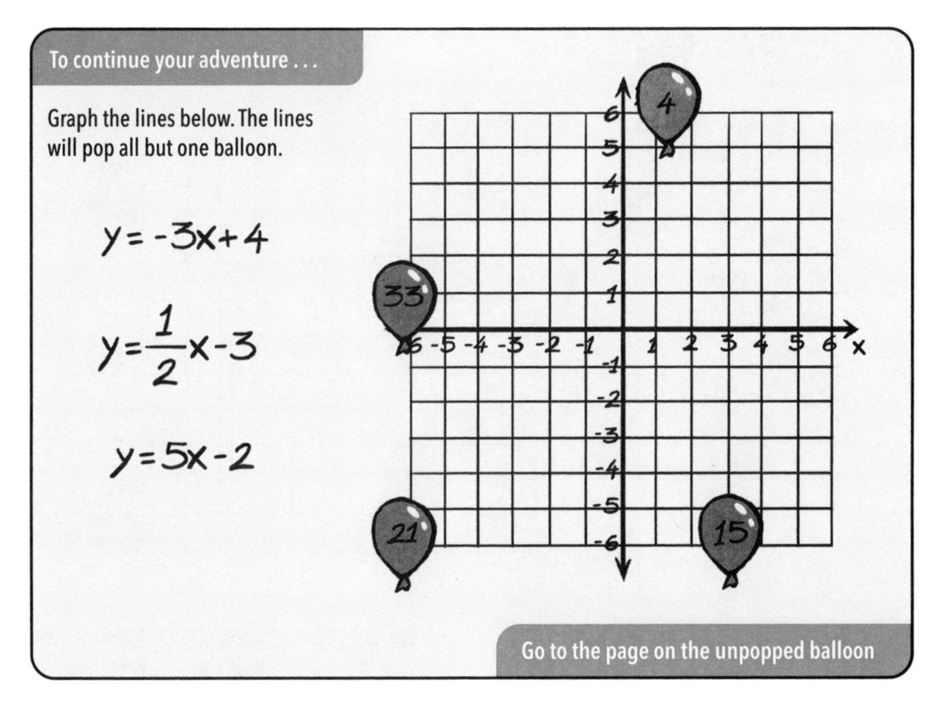

See page 54 of Adventurer's Advice Chapter 2 for help.

You should have arrived here from page 9

The wrench is bright red, and it's heavy. You rear back and swing it, as hard as you can, right at the nearest blue tube. The glass cracks, then the crack bulges out and the whole tube breaks open.

A boatload of thick blue goo comes glopping onto the floor of the lab. On that spilling slime, the sheep that had been floating inside the tube comes sliding out across the floor.

The sheep is wearing the same helmet-backpack apparatus as the dog. It shakes its head, then tries to stand in the slippery mess.

Without thinking, you help the sheep to stand. Once the sheep has collected itself, it turns to face the dog. Through the small speaker in its backpack, the sheep . . . speaks.

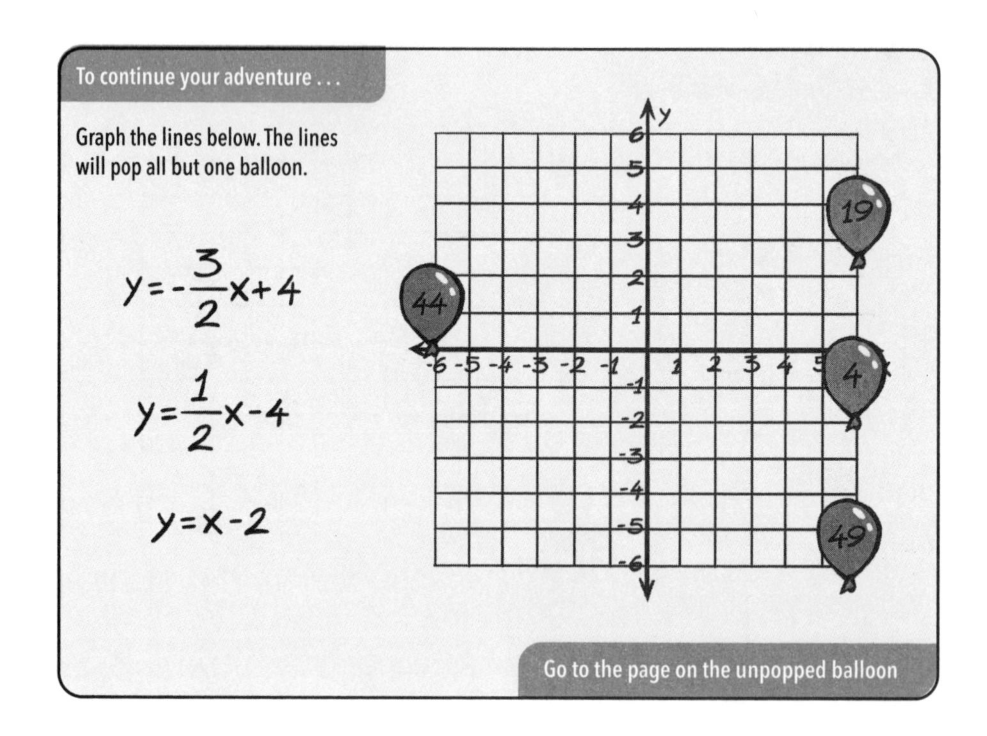

See page 54 of Adventurer's Advice Chapter 2 for help.

You should have arrived here from page 18

The knob on the generator has one more setting: "extra high voltage: extra danger." That seems kind of redundant . . . but you go ahead and crank it to that maximum level.

The room is now filled with smoke, orange flashing lights, and—from the row of robots—a chorus of "Er-ror! Er-ror!" It's the perfect cover to escape the robot from the stairs.

Through the smoke, you grope your way to the door. You push through and shut the door behind you.

You're in a long hallway now. Walking along it quickly, you come to an elevator. This could, you realize, take you at last to your goal—the eleventh floor.

You step inside.

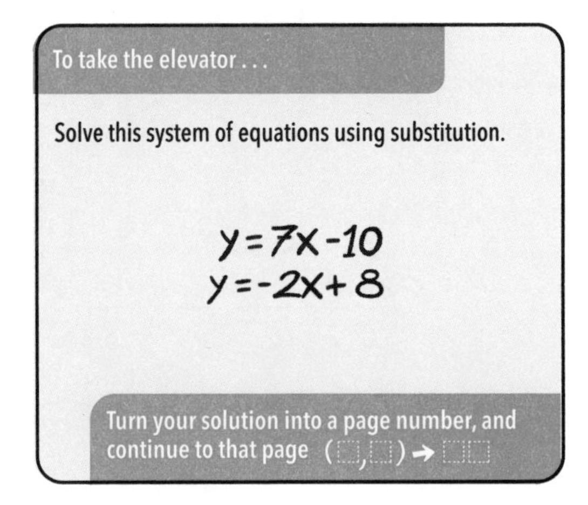

See page 57 of Adventurer's Advice Chapter 2 for help.

You should have arrived here from page 30

"And I better not hear another peep out of you, Bosco."

You round the corner, following that voice into what looks like a laboratory. The pig is in here scolding the Rottweiler, who's now sulking in the corner.

This lab is *really* weird. It's full of tall glass tubes, and each tube is filled with a glowing blue liquid. Floating in that liquid in each tube is an animal, wearing a metal helmet and black backpack like the Rottweiler and the pig.

You see a sheep. You see a dolphin. One extra-tall tube has a giraffe inside.

The pig turns to you. "Now, where were we?" it asks. "Right! If you follow that hallway there, you will find a service elevator. You can take that to wherever you need to go."

"Um . . . okay," you say to the pig. "So . . . thanks. I guess."

You leave the lab and walk down the long hallway—and sure enough, there's a service elevator. Its doors are open. Hoping it can take you to the eleventh floor, or at least get you closer, you step inside.

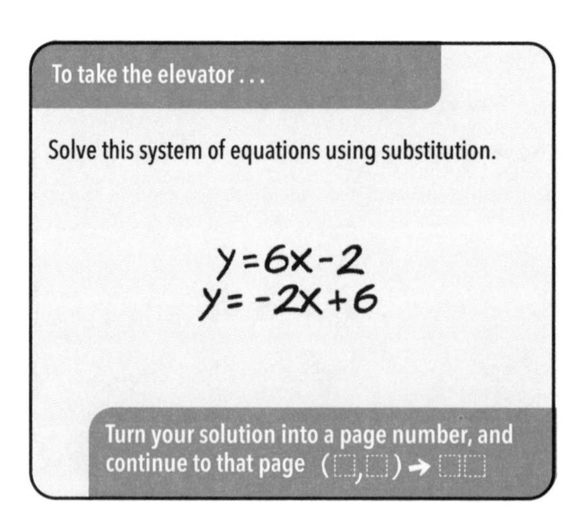

See page 57 of Adventurer's Advice Chapter 2 for help.

Go back to the previous page and check your work. You should not have arrived here.

You should have arrived here from page 35

The sprinklers have soaked you too, but you're much better off than the tangle of robots. Two more have slipped and fallen to the ground. A chorus of beeping, booping, and buzzing fills the room as the robots short-circuit from the water and begin to malfunction. One limps around in circles as its camera head emits a geyser of sparks.

You'd better get out of here before any functioning robots—or humans—show up. You run back to the hallway, leaving the robotic ruckus behind.

Striding quickly down the hall, you come to a service elevator. Its door is open, and it's empty. Almost as if it was waiting for you.

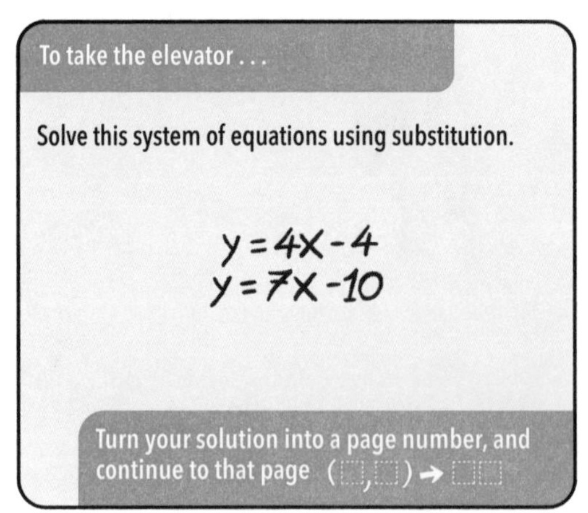

See page 57 of Adventurer's Advice Chapter 2 for help.

Visual patterns (page 1)

This problem gives us a pattern and asks for the 8th figure in the pattern. There are a bunch of strategies you can use to solve this question, and I hope you tried it in your own way!

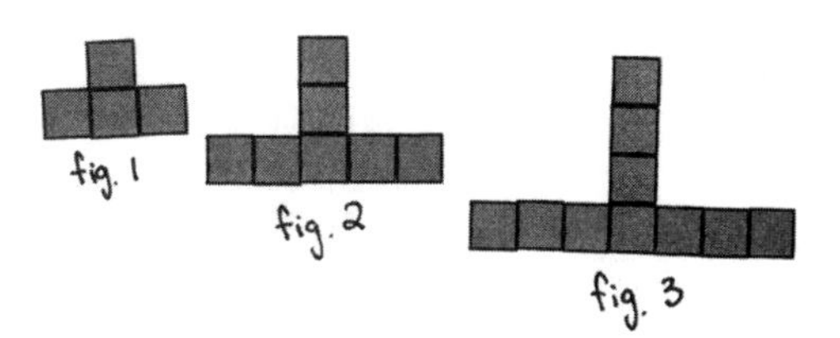

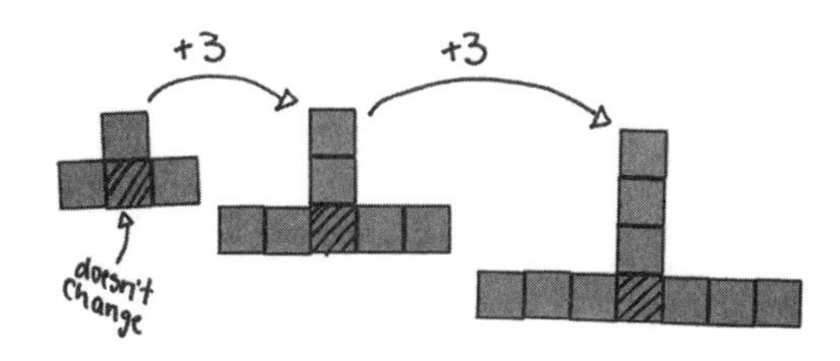

STEP ONE: Identify the pattern. The three shapes that we were given have 4, 7, and 10 blocks. This pattern has a constant rate of change. It's adding 3 blocks each time! (One way to solve this is to just keep adding 3 over and over until you get to figure 8!) We also have that one block in the middle of our figure that isn't changing between figures.

STEP TWO: Model the pattern with an equation. Adding patterns can be modeled with $y = mx + b$ equations. b (the y-intercept) is the starting value. In this problem, the starting value is the *one* block that we start with in the middle of each shape. m (the slope) is the rate of change. In this problem, we are adding *three* blocks each time, so the slope is 3. $y = 3x + 1$ will model this pattern!

$y = mx + b$ ← starting value (1)
↑ Rate of change (+3)

$y = 3x + 1$

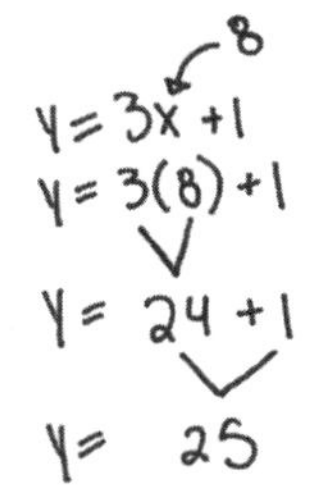

STEP THREE: Plug in 8 for x and solve. If we plug 8 (the figure that we're looking for) into the equation, it will tell us how many blocks we need. Follow order of operations and we get 25 blocks! Continue to that page.

ANSWER KEY FOR THIS TYPE OF PROBLEM:
Page 1: Continue to page 25

Slope from two points (pages 23 and 34)

This problem is asking about slope: basically, how *steep* a line is. It gives us two points that fall on that line and asks us to figure out: is this line really steep (is it going up quickly) or is it growing more slowly?

$(-5,-8)$ and $(-1,4)$

$(-5,-8)$ and $(-1,4)$ — x, y, x, y — -8 and 4: This distance (Ryse) is $+12$!

STEP ONE: Find the rise (aka "ryse"). I like to misspell "rise" as "ryse" to remind myself that the rise is the change in the y-direction. Rise tells us how far the line is going up (or down) between these two points. In our problem, we are going from −8 all the way up to 4. That distance is 12 spots, and we are going up from *negative* 8 to *positive* 4. We can say our rise (or ryse) is 12.

STEP TWO: Find the run. We know our line is going up 12, but over how long of a distance? We need to figure out how fast our line is climbing those 12 spots, and for that we find run. In this problem, we are going from −5 to −1, which is a distance of 4, and we are going up. This means our run is 4.

$(-5,-8)$ and $(-1,4)$ — x, y, x, y — -5 and -1: This distance (Run) is $+4$!

$$\text{Slope} = \frac{\text{Ryse}}{\text{Run}} = \frac{12}{4}$$

STEP THREE: Find the slope. Slope is a ratio. You can think of it as comparing change in y (ryse) to change in x (run). It compares how fast our line is going up (or down) with how fast it is going side to side. Our problem is going up 12 for every 4 spots it moves to the right, meaning our ryse/run ratio would be 12/4.

STEP FOUR: Reduce our slope. This ratio can be reduced. 12/4 is the same thing as 3/1 (it goes up 3 for every 1 space on the graph). 3/1 is the same as 3. Continue to that page!

$$\frac{12}{4} = \frac{3}{1} = 3$$

AUTHOR'S NOTE: You can also solve this type of problem using a graph! There is a more detailed explanation of this strategy on page 61!

ANSWER KEY FOR THIS TYPE OF PROBLEM:

Page 23 (left): Continue to page 7
Page 23 (right): Continue to page 10
Page 34 (left): Continue to page 3
Page 34 (right): Continue to page 12

y-intercept from two points (pages 3, 5, and 7)

This problem gives us two points on a line and asks us to find where that line starts (where that line crosses the y-axis).

(3,1) and (4,-3)

(3,1) (4,-3)
x y x y

Ryse: 1 to -3 Run: 3 to 4

Down 4 Up 1

$$\text{Slope} = \frac{\text{Ryse}}{\text{Run}} = \frac{-4}{1} = -4$$

STEP ONE: Find the slope. The first thing that we need to do is find the slope between these two points. Slope will tell us how steep our line is, and for these two points our line is going down 4 for every 1 that it goes side to side, meaning the slope is −4. (There is a much more detailed explanation of slope on page 52 of this chapter's Adventurer's Advice.)

STEP TWO: Plug what we know into the equation for a line. This line has a slope of −4. Where does a line that has a steepness of −4 need to start in order to go through (3,1)? Or (4,−3) for that matter? It helps us to know that the equation for any line can be written as $y = mx + b$. The m in this equation is the slope (−4 in this problem) and the b is the y-intercept (what we are looking for). The x and y values in our equation are just the x and y values for any point along the line. We have two points on this line to choose from, and I'm going to pick (3,1). When I plug in m, x, and y, I get a $y = mx + b$ equation that says $1 = (3)(-4) + b$.

y value, slope, x value: $y = mx + b$

Slope = −4 (3,1)
x y

$y = mx + b$
$(1) = (-4)(3) + b$

$(1) = (-4)(3) + b$
$1 = -12 + b$
$+12 \quad +12$
$13 = b$

STEP THREE: Solve for b. 3 times −4 is −12, and we can get the variable b by itself by adding 12 to both sides of our equation. We end up with $b = 13$, which means that a line that goes through the two points we were given has to start at 13. Continue to that page!

ANSWER KEY FOR THIS TYPE OF PROBLEM:

Page 3 (left): Continue to page 13
Page 3 (right): Continue to page 9
Page 5 (left): Continue to page 6
Page 5 (right): Continue to page 2
Page 7 (left): Continue to page 11
Page 7 (right): Continue to page 8

Graphing lines (pages 17, 18, 30, 35, 39, 45, and 46)

This problem is asking us to graph some lines, and all the lines are given in y = mx + b format. Let's graph y = −2x + 3.

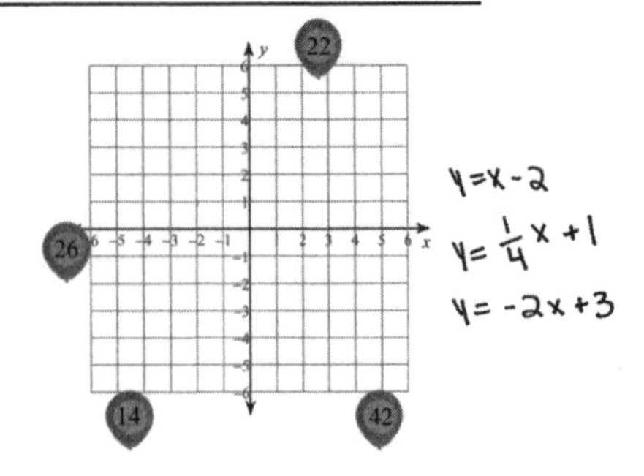

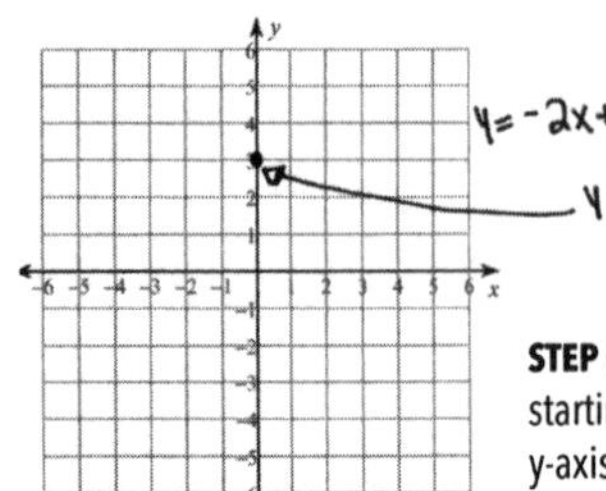

STEP ONE: Plot your y-intercept. The y-intercept on a graph is the starting point, and it's where our line crosses (or *intercepts*) the y-axis. This line needs a y-intercept at 3.

STEP TWO: Use your slope to plot the next point. This slope has a slope of −2, which means that it goes down 2 for every 1 space that it goes over. It can help to think of −2 as a fraction. −2/1 is the exact same number as −2. Count down 2 spaces, over 1, and place your next point.

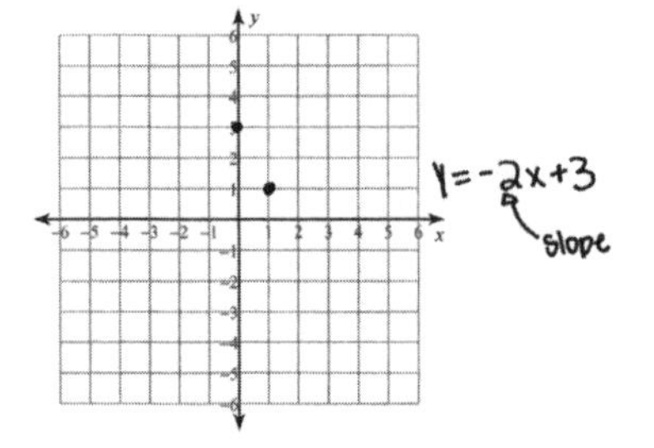

STEP THREE: Continue your pattern. Continue following your slope pattern (down 2, right 1) until your line is complete. This line pops the 42 balloon, meaning that we *don't* want to go to that page.

STEP FOUR: Graph the other lines. Follow Steps One through Three with the other two lines. These lines will pop the 14 and 26 balloons, leaving the 22 balloon as the only one left. Continue to that page!

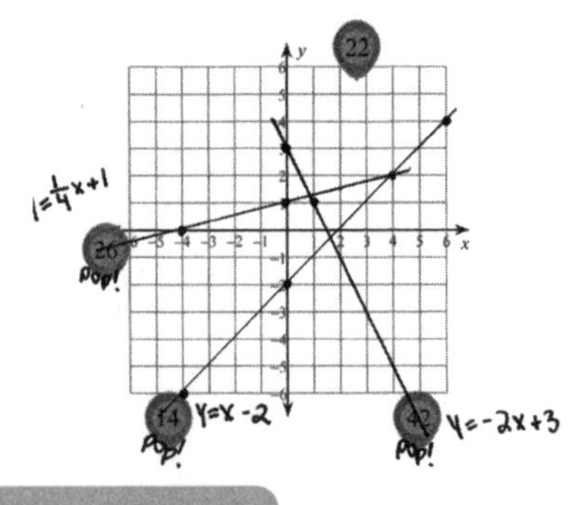

ANSWER KEY FOR THIS TYPE OF PROBLEM:

Page 17: Continue to page 22
Page 18: Continue to page 47
Page 30: Continue to page 48
Page 35: Continue to page 50
Page 39: Continue to page 41
Page 45: Continue to page 33
Page 46: Continue to page 44

Visual system of equations (pages 14 and 24)

This question is sort of like the other systems of equations problems that we have looked at in this chapter, but this one uses barnyard animals instead of variables like x, y, or z.

sheep + pig + sheep = 19

2 chicken = sheep

chicken + chicken + chicken = 9

To continue your adventure, find...

sheep + chicken + pig

And continue to the page that matches your answer.

This could be 2, 3, 4, 10, -100, 0...

This equation we can solve!

STEP ONE: Find a place to start. The first equation with the sheep and the pig will help us later, but right now it has too many variables. Until we get some more information, the sheep could be any number! If we keep looking, the third equation only has chickens. This is an equation that we can solve right now!

STEP TWO: Solve for the first variable. The third equation is a great place to start, and if we know that three chickens are worth 9, each chicken must be worth 3, since 3 + 3 + 3 = 9..

STEP THREE: Solve for the second variable. We know that the chicken is worth 3 now, and that will help us in the second equation! The second equation tells us that a sheep is worth 2 chickens. We know what a chicken is worth, so the sheep must equal 6, since 2 * 3 = 6.

STEP FOUR: Solve for the third variable. If the sheep is 6, then we can use the first equation to solve for the pig! Each one of the sheep is worth 6. That means that the pig must equal 7 so that we get to 19 (6 + 7 + 6 = 19). Now we have all three variables figured out!

STEP FIVE: Find the page! We know the pig, the sheep, and the chicken now, but this question is asking what sheep + chicken + pig equals. Substitute in the values that we found in Steps Two, Three, and Four, and we can see that we should continue to page 16!

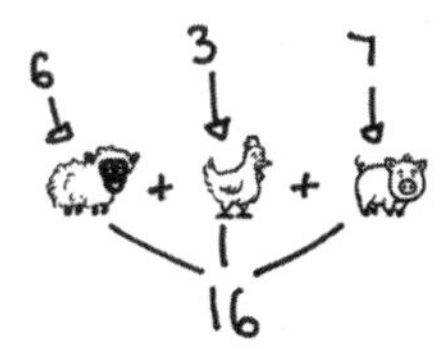

ANSWER KEY FOR THIS TYPE OF PROBLEM:

Page 14: Continue to page 16
Page 24: Continue to page 16

Solving systems by graphing (pages 32 and 38)

This problem is looking for the *solution* to this system of equations. The solution is the place where the two equations are exactly equal, and on a graph, that point is where the lines cross.

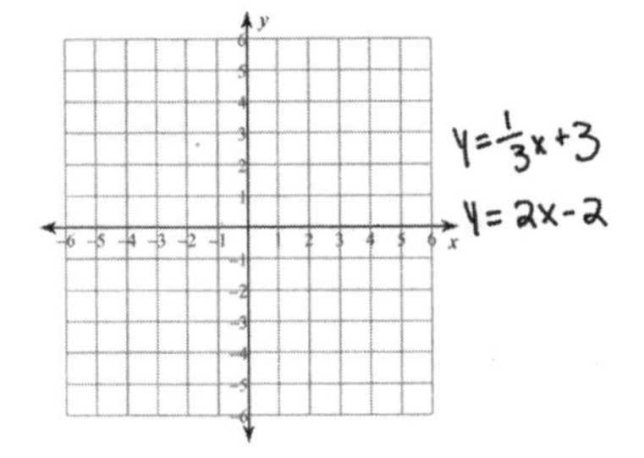

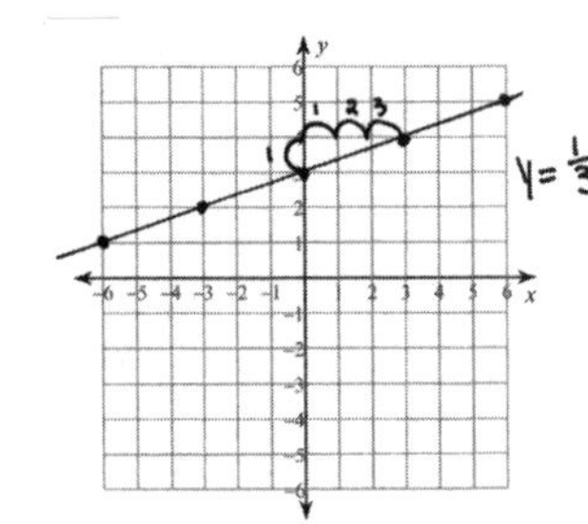

STEP ONE: Graph the first line. y = 1/3x + 3 starts at 3 and goes up 1 spot for every 3 it goes to the right. There is a much more detailed explanation for how to graph lines on page 54 of this chapter's Adventurer's Advice.

STEP TWO: Repeat with the second line.

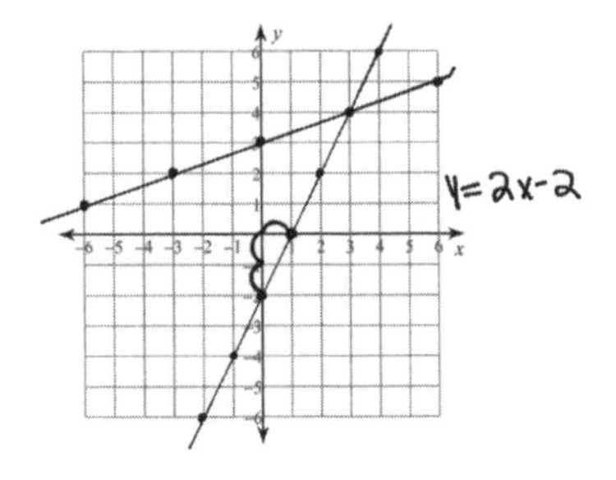

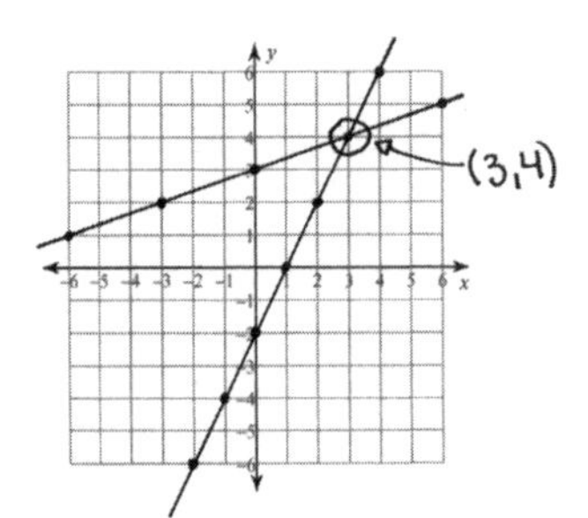

STEP THREE: Find the place where the two lines cross. In this problem, our two lines cross at the point (3,4). If we turn that point into a page number, we get 34, so we will continue to that page.

ANSWER KEY FOR THIS TYPE OF PROBLEM:

Page 32: Continue to page 34
Page 38: Continue to page 23

Systems and substitution (pages 22, 33, 41, 44, 47, 48, and 50)

This problem is looking for a solution to this system. The solution is the place where these lines cross and these equations are equal. Before we start, it helps to remember that the places these lines are going to cross is going to be an (x,y) ordered pair, so we need to find a solution in that format. For this problem, both equations are solved for y, which makes it easy to solve with a strategy called *substitution*.

$$y = 6x - 8$$
$$y = 8x - 12$$

$(\square,\square) \rightarrow \square\square$

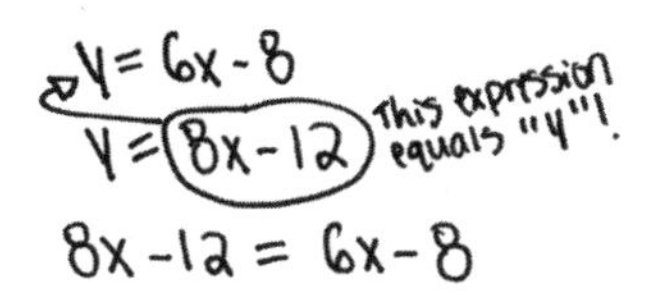

STEP ONE: Substitute. The second equation tells us that y is equal to 8x – 12. Those two quantities are exactly the same, so we can look to the first equation. Instead of y in the first equation, we can write in the thing that y is equal to: 8x – 12. This gives us the new equation 8x – 12 = 6x – 8. This helps us because now we have only one variable and we can solve!

STEP TWO: Solve for x. Take away 6x on both sides, then add 12 to both sides, before dividing both sides by 2. (There is a more detailed breakdown of this solving strategy on page 51 of Chapter 1 Adventurer's Advice.) We find that the place the lines cross happens when x is equal to 2. This isn't our final solution, but we do know that the x in our (x,y) ordered pair is going to be 2.

$$8x - 12 = 6x - 8$$
$$-6x \quad -6x$$
$$2x - 12 = -8$$
$$+12 \quad +12$$
$$2x = 4$$
$$\div 2 \quad \div 2$$
$$x = 2$$

$$x = 2$$
$$y = 6x - 8$$
$$y = 6(2) - 8$$
$$y = 12 - 8$$
$$y = 4$$

STEP THREE: Plug in our x-value and solve for y. From here, we need to find where our lines are when x is equal to 2. We do that by plugging in 2 for x. It doesn't actually matter which equation we pick, because this is the *one place* where the two lines are at the same place on our graph! I'm going to use y = 6x – 8, but you can use the other equation if you want. Follow order of operations, and we find that if we multiply 6 by 2 then take away 8, we're left with a y-value of 4.

STEP FOUR: Put your x and y values together to find the solution. We have already done all the math, now we just need to plug 2 and 4 into an (x,y) ordered pair. Step Two told us that x = 2, and Step Three told us that y = 4, so the solution is (2,4). Turn that point into a page number (24) and continue to that page!

$$x = 2 \quad y = 4$$
$$(x, y)$$
$$(2, 4)$$

ANSWER KEY FOR THIS TYPE OF PROBLEM:

Page 22: Continue to page 14
Page 33: Continue to page 24
Page 41: Continue to page 14
Page 44: Continue to page 24
Page 47: Continue to page 24
Page 48: Continue to page 14
Page 50: Continue to page 24

Systems and elimination (pages 2, 6, 8, 9, 11, and 13)

This problem is looking for a solution to this system. The solution is the place where these lines cross and these equations are equal. Before we start, it helps to remember that the places these lines are going to cross is going to be an (x,y) ordered pair, so we need to find a solution in that format. For this problem, neither equation is solved for x or y, which means substitution would be difficult. Instead, we're going to use a strategy called *elimination*.

$$4x + 4y = 28$$
$$-2x + 6y = 10$$

(□,□) → □□

$4x + 4y = 28$ (equation 1)

$-2x + 6y = 10$ (equation 2)

pick this one!

$\cdot 2 \rightarrow -4x + 12y = 20$

STEP ONE: Look for a way to make one of our variables cancel (add to zero). In Step Two, we're going to make one of our variables cancel out. To do that we need equal and opposite numbers of x's or y's in the two equations. (For example, *positive* 5y in the first equation with a *negative* 5y in the second equation. Those terms are equal and opposite, and would cancel out.) In our problem, if I double my second equation, I will get a "negative 4x," and that will cancel out with the "positive 4x" in our first equation. When we multiply the *entire* second equation by two, we get a new equation: $-4x + 12y = 20$.

STEP TWO: Stack the equations and combine the like terms, now that we know one variable will cancel. Using our new, doubled version of equation 2 and the original version of equation 1, we can add both equations together. The positive 4x and the negative 4x cancel (that's why we multiplied the second equation by two) and leave us with $16y = 48$. Divide both sides by 16, and we get $y = 3$. This isn't our final solution, but it tells us that when these two lines cross, they are at a height of 3. The y in our (x,y) point will be 3.

$4x + 4y = 28$ (original equation #1)
$-4x + 12y = 20$ (new equation #2)

ADD

$0x + 16y = 48$
$16y = 48$
$\div 16 \quad \div 16$
$y = 3$

$y = 3$
$-2x + 6y = 10$
$-2x + 6(3) = 10$
$-2x + 18 = 10$
$-18 \quad -18$
$-2x = -8$
$\div -2 \quad \div -2$
$x = 4$

STEP THREE: Plug our value into one of our original equations to find the other variable. From here, we need to find where our lines are when y is equal to 3. We do that by plugging in 3 for y. It doesn't actually matter which equation we pick, because this is the *one place* where the two lines are at the same place on our graph. I'm going to pick $-2x + 6y = 10$, but you can use the other equation if you want. After we plug in 3, we can solve and find that when $y = 3$, $x = 4$ (for a detailed explanation of this solving strategy, go to page 51 in Chapter 1 Adventurer's Advice).

STEP FOUR: Put your x and y values together to find the solution. We have already done all the math, now we just need to plug 3 and 4 into an (x,y) ordered pair. Step Two told us that $y = 3$, and Step Three told us that $x = 4$, so the solution is (4,3). Turn that point into a page number (43) and continue to that page!

$x = 4, \quad y = 3$
(x, y)
$(4, 3)$

ANSWER KEY FOR THIS TYPE OF PROBLEM:

Page 2: Continue to page 43
Page 6: Continue to page 43
Page 8 (left): Continue to page 35
Page 8 (right): Continue to page 45
Page 9 (left): Continue to page 27
Page 9 (right): Continue to page 46
Page 11 (left): Continue to page 18
Page 11 (right): Continue to page 36
Page 13 (left): Continue to page 31
Page 13 (right): Continue to page 17

Graphing linear inequalities (page 25)

Most of the other problems in this chapter ask us to graph an equation where y *is equal* to some version of x. This problem is different. It's an *inequality* and there are a lot more places where y is *unequal* to some version of x. We read "≤" as "less than or equal to," so this problem in particular is asking us to show all the places on the graph where y is *less than OR equal to* 3/2x – 5.

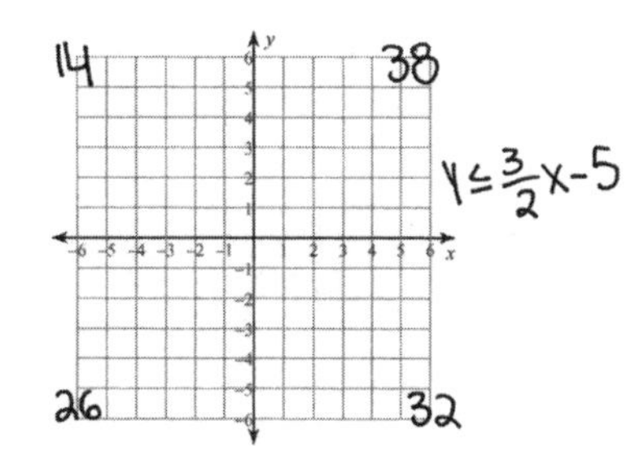

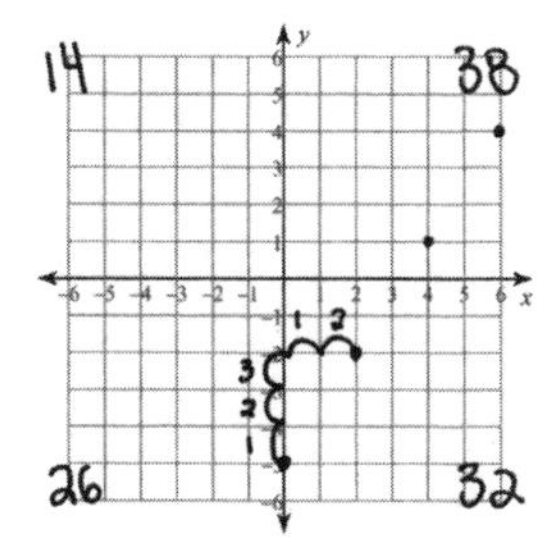

STEP ONE: Graph the line, same as normal. Check page 54 of this chapter's Adventurer's Advice for a more detailed breakdown, but we start this line with a y-intercept of negative 5, and the slope tells us to go "up 3, over 2" to create a linear pattern. Plot all the points that follow this pattern. This line is the "3/2x – 5" part of our inequality.

STEP TWO: Decide if the line is solid or dotted. A solid line means that we are *including* the points on the line. A dotted line means that we are *not including* the points on the line. In this problem, we are graphing all the places on the graph where y is less than *or equal to* the line. That "or equal to" part means that we are including the points on the line, so this line would be solid.

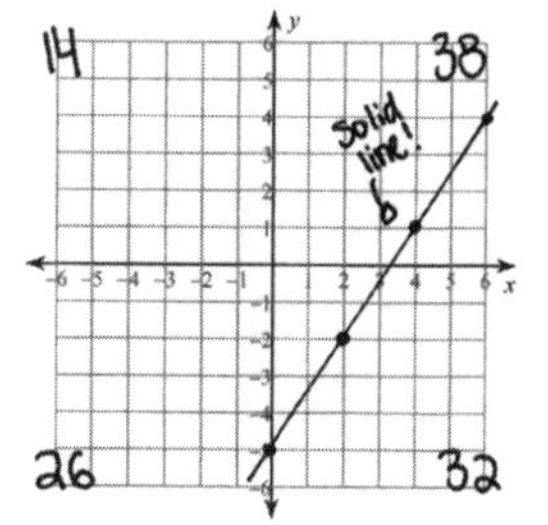

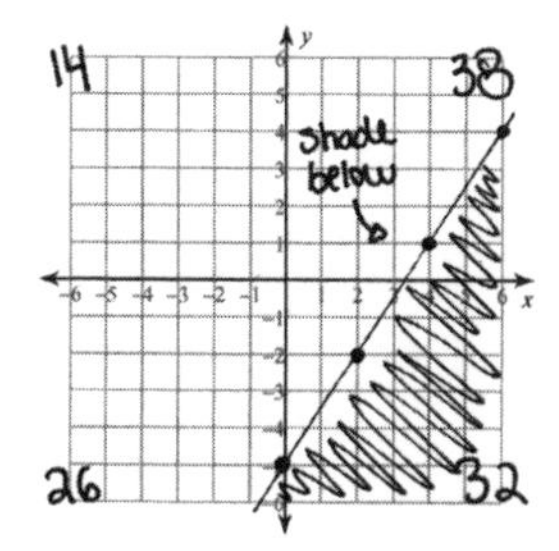

STEP THREE: Shade. This problem wants us to show all the places that y is *less than* or equal to the line. The "less than" part means we want to shade "less than" or below our line. The only page number in this shaded portion of our graph (called the "solution set") is 32, so continue to that page!

ANSWER KEY FOR THIS TYPE OF PROBLEM:

Page 25 (top): Continue to page 38
Page 25 (bottom): Continue to page 32

Coordinate pairs

Need some help plotting points on a graph? Even though this skill isn't in *this* book, it's a super important one! Coordinate pairs (aka points) are always written the same way: (x,y). x first, then y. The order *matters* here and we need to be careful to plot our points correctly! Remember, x comes before y, just like in the alphabet.

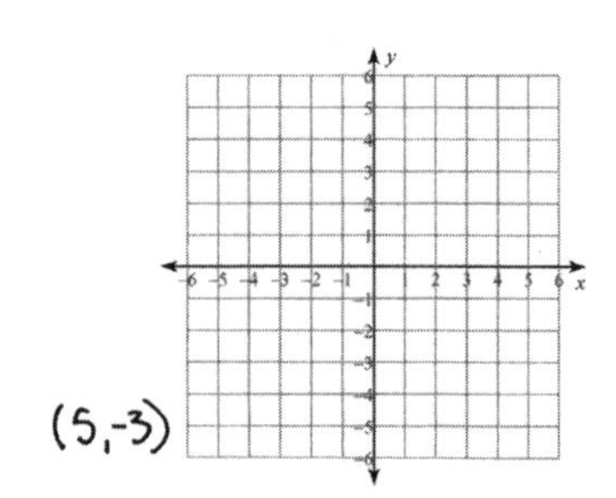

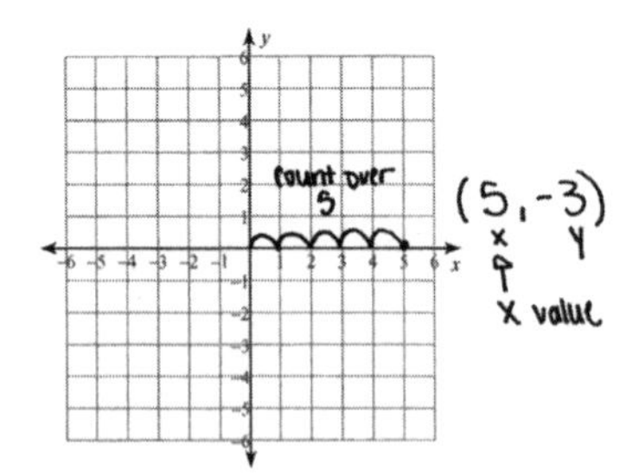

STEP ONE: Move side to side to find your x value. If we want to plot (5,−3), the first thing that we do is count to 5 along the x axis. x comes first in our coordinate pair, and the x-axis is the horizontal axis, so we count 5 spots to the right from the origin of our graph.

STEP TWO: Move up or down to find our y value. We already counted over to the right 5, and now we look to our y-coordinate. To plot the point (5,−3), that y-value will be −3, which means we count *down* 3 and plot our point.

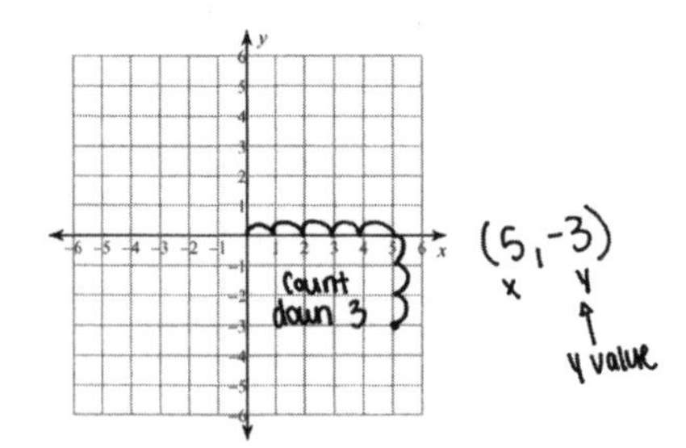

Slope from a graph

Want some help finding slope? This page shows one possible strategy: Using a graph!

(-5,-8) and (-1,4)

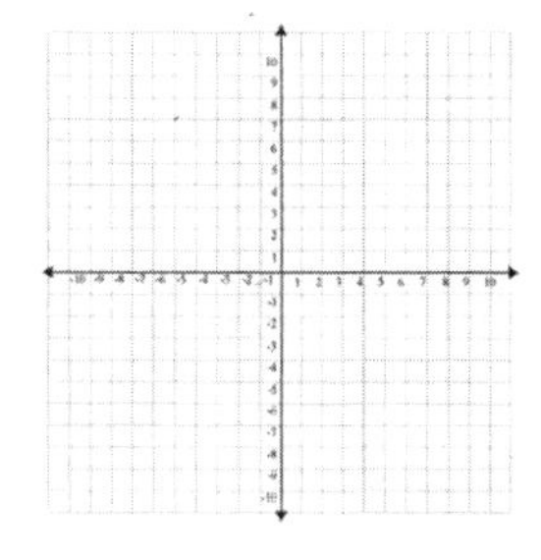

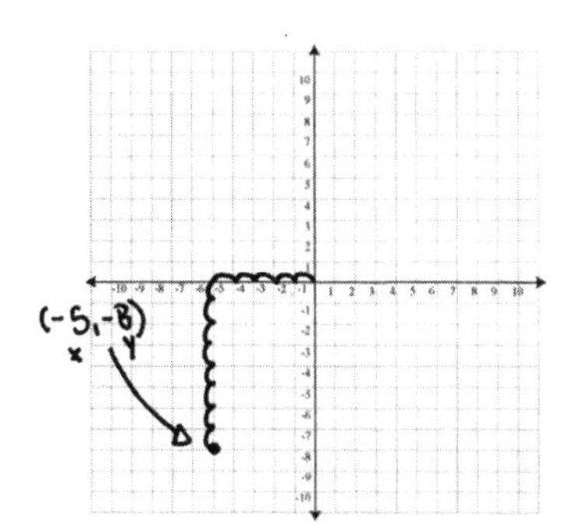

STEP ONE: Plot the first point. There is a more detailed explanation on page 54 of this chapter's Adventurer's Advice, but if we want to plot (−5,−8), we want to count to −5 on the x-axis. This means we count left 5 places from the origin to find −5. From there, we count down 8 places on the y and place a point there.

STEP TWO: Repeat Step One with the second point, this time counting left 1 before moving up 4 places. I like to connect the points and draw a triangle to help in Step Three.

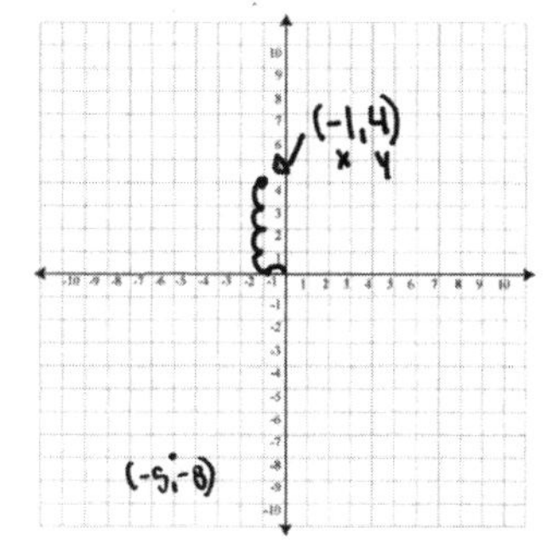

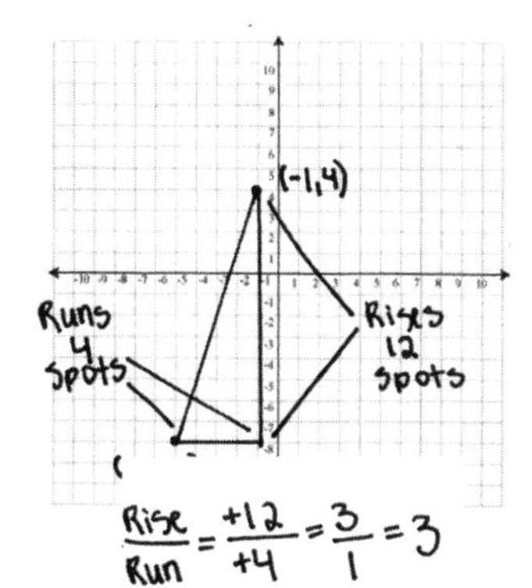

STEP THREE: Find the slope. Slope tells us how steep a line is, and it helps to think of slope as *rise over run*. Now that we have graphed 2 of the points that land on that line, we can look at the steepness (aka slope). *This line would go over 4 spots on the graph for every 12 spots that it goes up.* Slope lets us put that idea into the ratio 12/4. We can reduce that ratio to 3/1, or just 3. (If you look closely at the graph, you will see some other points along the line in an "up three and over 1" pattern.)

AUTHOR'S NOTE: Graphing to find slope works great, but sometimes we have numbers that are just too big to use a graph. If you run into that type of problem, you may want to try the steps on page 52 of this chapter's Adventurer's Advice to find slope.

CHAPTER 3

You step off the elevator and onto the eleventh floor. At last! You've reached Doctor LaBella's lab. You can return her wallet and ID, collect your reward, and get OUT of this wild place.

Ahead of you are metal swinging doors leading straight to Doctor LaBella's lab. You push and pull against the doors, but they don't budge. They must be locked with a dead bolt. You lift Doctor LaBella's ID to the door and try to scan yourself in, but nothing happens. There are small rectangular windows in each door, and you peer through them to the lab inside.

The lights are off, but something is glowing bright blue and casting swirling shadows against the walls of the lab.

You pull yourself away from the little window in the door. You're not getting in without a key. The only other room on this floor is across the hall. It has a narrow wooden door with simple, black lettering that says **NON-HUMAN RESOURCES.** You knock on the door. Maybe someone inside can get Doctor LaBella for you.

No answer.

You check the handle and it's unlocked.

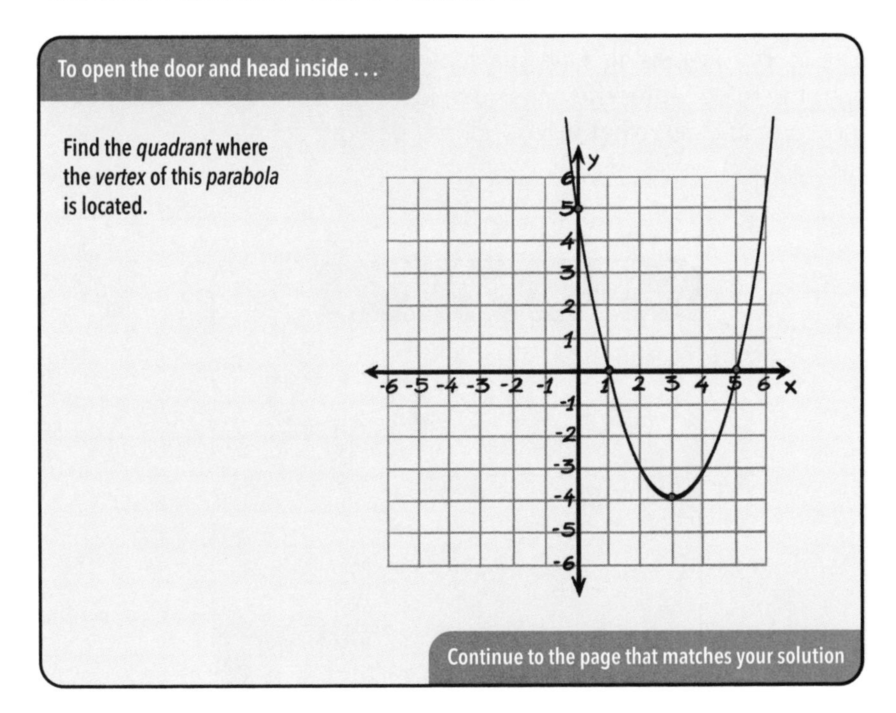

See page 51 of Adventurer's Advice Chapter 3 for help.

You should have arrived here from page 29

Your reasoning is simple. The cardboard of the boxes should be easier to drive through than the metal of the robots. The speed on the forklift is maxed out, and you yank the steering wheel to the left as hard as you can. The long, pointy fork at the front of your vehicle crashes into a ten-foot-high tower of boxes, tearing through the cardboard and knocking the whole thing over. The boxes release an avalanche of tiny glass vials that scatter all over the cement floor.

To your dismay, the little forklift wheels are no match for the cellophane wrapping the boxes. You punch the gas, but you aren't going anywhere. Without the cover and the speed of the forklift, you don't stand a chance against the robots. You dive out of the cab to the floor and look around frantically.

Twenty feet away, the robots are scurrying toward you and will be there any second. Lucky for you, their skinny metal legs were not designed to handle such treacherous footing. They slip on the spilled insulin vials and crash to the ground.

Two other Dregg robots come quickly to see what caused the massive spill, but they are too late. You are able to slip down a corridor of boxes and past them undetected.

The portal is pulsing blue plasma, and you run as fast as you can toward the thing. After the chaos you just caused, you don't want to stick around and find out what will happen if you get caught.

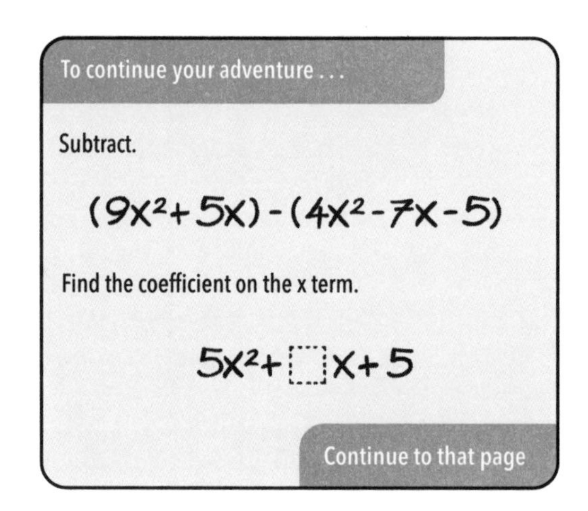

See page 53 of Adventurer's Advice Chapter 3 for help.

Go back to the previous page and check your work. You should not have arrived here.

You should have arrived here from page 1

You twist the handle to the narrow wooden door, and it squeaks as it opens. The room inside is spacious, with small TV screens covering most of the wall space. There must be hundreds of them. *Who needs this many TVs?* There aren't any people in here, or any other doors out of this room, so this might be a dead end. The screens are playing security camera footage, and you take a step closer to investigate.

The closest TV is showing a grizzly bear onscreen with a white stripe on its back, loading cardboard boxes onto a dolly. Any other day, you would have laughed at such a preposterous sight, but after the evening you have had so far, you know that Dregg is capable of anything. A small strip of tape above the screen reads: "Non-Human #517." Does Dregg have these animals doing manual labor?

There is only one button on the TV and you press it with your thumb. Maybe something else is on? The channel doesn't change, but the bear onscreen twitches fiercely, as if it is being hit with a taser. You look more closely at the button, and to your horror you realize that small letters beside the button spell out "SHOCK COLLAR" in all caps.

You just sent this bear a jolt of electricity from here in this small, strange room! He glares up at the camera and growls at you, but he returns to loading boxes on the cart. He doesn't want to get shocked again.

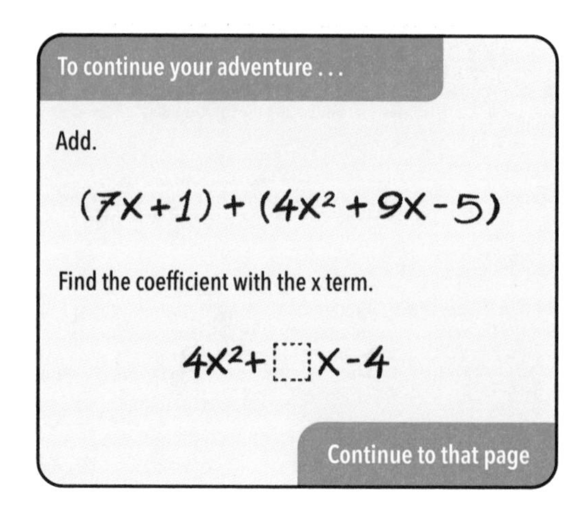

See page 52 of Adventurer's Advice Chapter 3 for help.

You *have to* cancel the data download. There is too much money on the line. You take a deep breath and click the “cancel download” button.

A pop-up box appears: “Are you sure?” You click “OK.”

Another pop-up box appears: “Are you *sure* you’re sure?”

You click “OK” again.

Another pop-up box. “OK, *fine*. You’re no fun. Download canceled.”

You sneak away from the server room and back to the bank lobby. You hear a shout from behind you: “Who stopped the download!?”

You pick up your pace to a run: across the lobby and toward the portal.

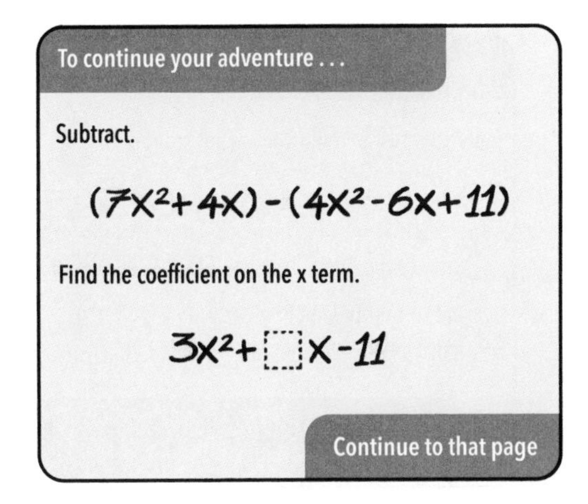

See page 53 of Adventurer's Advice Chapter 3 for help.

You should have arrived here from page 11

This bank is big. It must be one of the biggest banks in the city, and maybe the only one with a vault. The broad corridor leading to the vault is tiled with a teal and purple checkerboard pattern.

You notice a puddle of bright pink paint on the tile floor, and you look up to see where it is coming from. Ten feet up the wall is a security camera, but the lens has been painted over. The bright pink paint is dripping onto the floor. Some*thing* has stepped in the bright pink puddle and tracked paint down the hallway. *The footprints don't look human*.

You feel uneasy. You don't know what Dregg is doing here, but if it was legal, they wouldn't need to do it at night, and they wouldn't need to cover up the camera.

You continue down the hallway, ducking from doorway to doorway, in case you come across someone *or something* from the Dregg Corporation.

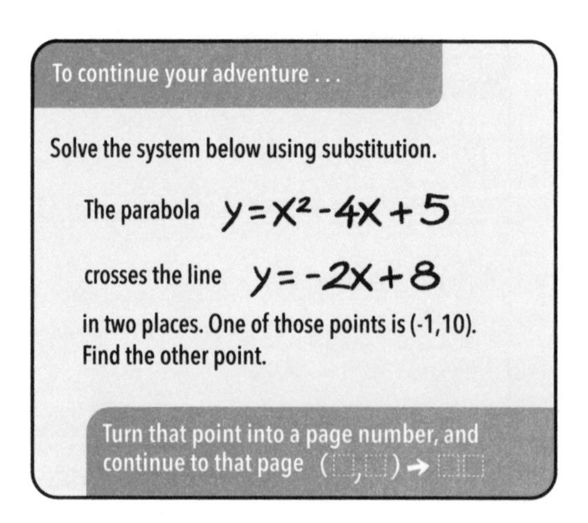

See page 65 of Adventurer's Advice Chapter 3 for help.

You should have arrived here from page 9

You sneak past rows of cardboard boxes emblazoned with the Dregg logo, and you quietly ascend a short set of stairs to the small office of the warehouse. There's not much inside. There are two desks, a mini-fridge in the corner, and a small window that opens out into the warehouse.

You ruffle through the papers on the desks, but nothing looks out of place. You peek through the window and can see four or five figures moving around on the warehouse floor. There are two bears that look just like the one that you shocked earlier. Same little backpack, probably the same shock collar. The others are robots with long angled legs and pointy heads. The bears seem to be loading boxes of insulin onto handcarts, and the robots are moving them toward the open portal. *Why would Dregg steal their own medicine?*

As you watch, one of the robots moving boxes on the floor loses the top box on its hand cart, and it falls to the floor, breaking open. Small glass vials clatter all over the smooth warehouse floor.

"What was that?" comes a voice from directly behind you . . .

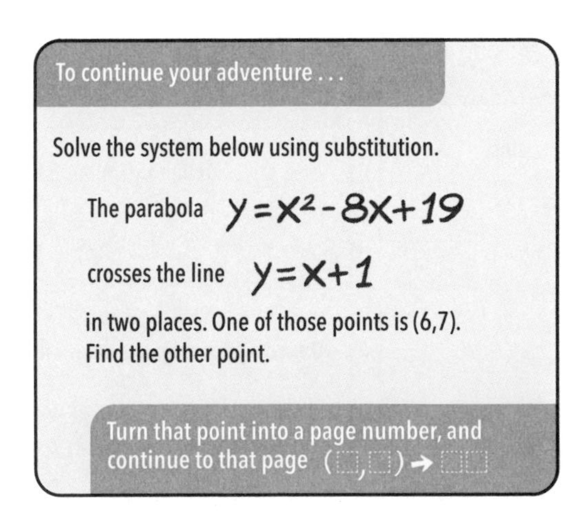

See page 65 of Adventurer's Advice Chapter 3 for help.

You should have arrived here from page 48

As you get closer, you realize you can see something *through* the blue coils of pulsing energy in the center assembly. It's almost like a little movie, but a really boring one. You can see rows of boxes and distant walls, but they are way taller than the walls of the lab. The space beyond the glowing energy is far larger than the space here in Dregg Tower. The boxes look just like the ones that you saw the bear loading onto a dolly. Is it possible that security camera footage was filmed *here*?

Without thinking, you dip your hand slowly into the blue plasma. It feels cold, and strange. Stranger still, your hand now appears to be on the other side of the plasma, playing a part in the movie that you are watching. Your hand is *in the warehouse* with the boxes.

Are these things . . . portals?

PORTALS! They can't be . . . How long has Dregg had this technology? Why have they never made this public? *Is this what Doctor LaBella was working on?* Just think of all the amazing things you could use this for!

You step close to the other blue plasma blob and peer in, wondering what this portal will show. Through the spinning blue energy on the right, you can make out a long table and some plants against the walls. It looks like a lobby, possibly to a bank.

You step over a series of huge cables connecting the steel frames of the portals to power outlets in the wall while you weigh your options. Two portals, two paths forward.

To take the center portal to the warehouse . . .

Solve this equation by isolating the variable.

$$x^2 + 20 = 101$$

Disregard the negative solution.

Continue to the page that matches the positive solution

To take the portal on the right to the lobby of the bank . . .

Solve this equation by isolating the variable.

$$x^2 - 20 = 101$$

Disregard the negative solution.

Continue to the page that matches the positive solution

See page 59 of Adventurer's Advice Chapter 3 for help.

You take a deep breath and leap into the swirling plasma of the center portal, landing in the warehouse on the other side. You can see back, through the portal, into the lab in Dregg Tower from your vantage on the warehouse floor.

This warehouse is *enormous*. Cardboard boxes are stacked ten feet high atop wooden pallets. They are arranged in neat rows on the smooth cement floor. The boxes look ready to ship, and you take a moment to read the label on a box nearby. It has a Dregg label, and it appears to be insulin. Dregg has been selling pharmaceuticals for years, and it seems that the portal from the lab brought you to a warehouse where they are shipping out insulin.

Each little vial of the stuff is pretty cheap to make, but Dregg overprices the stuff when they sell it. The inventory in all these boxes is worth a fortune!

What is Dregg up to here? This is *their* warehouse, and *their* medicine. There is an office up a set of stairs against the back wall. You could head up there, or you could continue to look for answers here on the warehouse floor.

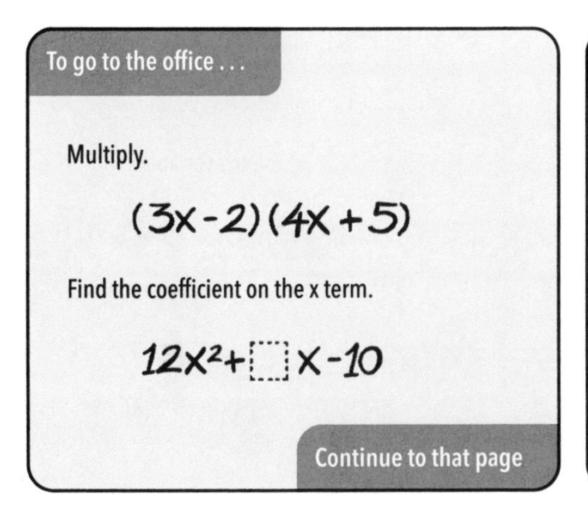

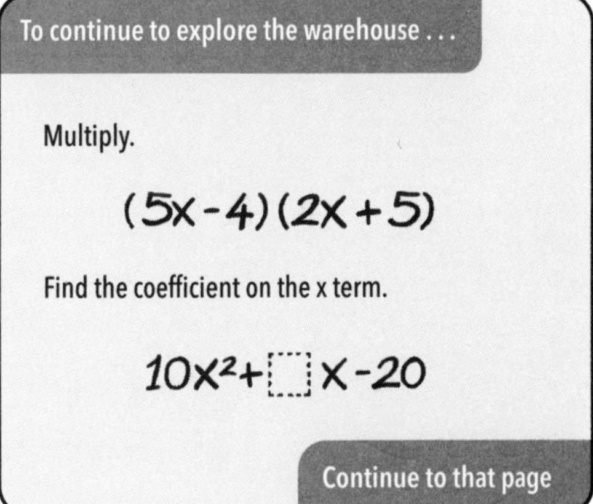

See page 54 of Adventurer's Advice Chapter 3 for help.

You should have arrived here from page 5 OR page 36

CHAPTER 3 GOAL: Achieved. You've uncovered some key Dregg secrets and learned how to travel through the portals.

You sprint through the bank lobby and leap through the swirling energy of the portal. The portal spits you out back in the lab on the eleventh floor of Dregg Tower.

You will have company soon, and you need to figure out a way to close off this portal behind you. There are three matching portal control panels, and you scan the right-most panel quickly for a way to shut it down. There is a big, blue OFF button and you jam it with your thumb.

The spinning blue plasma sputters a bit and dwindles until it has closed completely, stranding the bozos in the bank behind you. The portal in the center is still open and spinning, and you could go investigate the warehouse on the other side before continuing.

To go through the open center portal to the warehouse, turn to page 9.

To skip the center portal, continue reading below.

You decide the center portal isn't worth it. Whatever is going on in there is probably just as shady as what you saw at the bank.

You press the blue OFF button on the center console, and the portal in the center also sputters and fades.

Continue your adventure in Chapter 4.

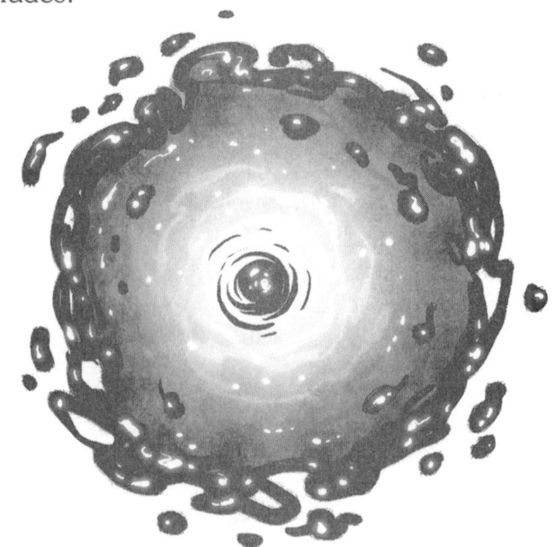

You take a deep breath and leap into the swirling plasma of the portal on the right. You land in the lobby on the other side. You look back where you came from, and you can see the lab in Dregg Tower.

You peer around. This is definitely a bank, but it is dark and empty. You see signs saying **MYRIAD CAPITAL BANK** near the stalls where the bank tellers would be during work hours. You know that banks move money to the back at night, and if Dregg is up to anything fishy, that's probably where you should look.

You sneak down the hallway past the lobby, and before long the hallway splits. A sign on the wall explains your options from here. To the left, a hallway leads to the server room. To the right, a hallway leads to the bank vault.

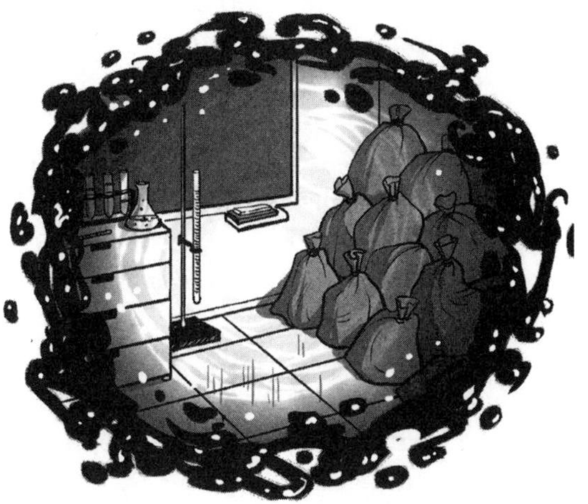

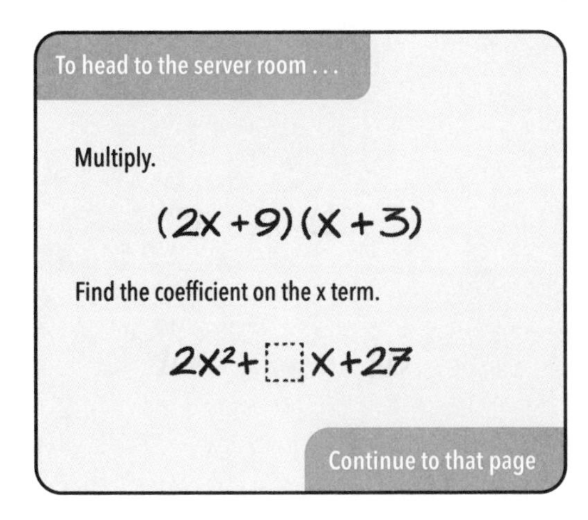

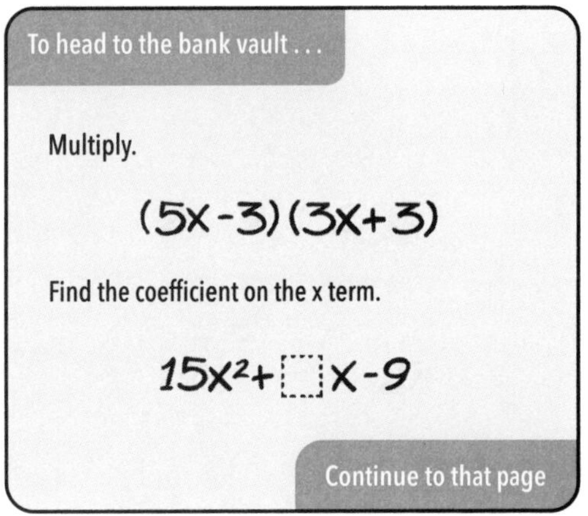

See page 54 of Adventurer's Advice Chapter 3 for help.

You should have arrived here from page 2, 22, 30, 40, OR 45

CHAPTER 3 GOAL: **Achieved. You've uncovered some key Dregg secrets and learned how to travel through the portals.**

The chaos that you caused on the warehouse floor was enough cover for you to escape. You scamper through the warehouse and leap into the spiraling blue ooze. You land with a thud back in the lab on the eleventh floor of Dregg Tower.

Immediately you turn to the control panel.

There are three matching control panels, one for each portal, and you press the OFF button on the panel that corresponds to the portal you just came through. The spinning blue plasma sputters a bit and closes, stranding the goons in the warehouse behind you.

The portal on the right is still open and spinning, and you could go investigate the bank lobby on the other side before continuing to Chapter 4.

To go through the open portal to the bank lobby, turn to page 11.

Otherwise . . .

You decide the second portal isn't worth it. Whatever is going on in there is probably just as shady as what you saw at the warehouse.

You press the blue OFF button on the right console, and the portal in the center also sputters and fades.

Continue your adventure in Chapter 4.

You should have arrived here from page 32

You nose through the duffel bag a bit. The bag is mostly empty, except for a roll of duct tape and six or seven bananas. *Weird.* Suddenly, the radio crackles to life. You pick it up and listen to the voice on the other end.

"Alpha unit, this is Beta unit. Status on the vault?" the voice says through the static of the radio.

Beta unit? What a stupid name. These goons aren't even original.

You press the button and muffle your voice through the collar of your shirt. "Beta, this is Alpha. Some heavy stuff down here. We could really use your help."

You duck into a darkened office across from the vault and wait for Beta unit.

The radio crackles again. "Roger, Alpha unit. On our way."

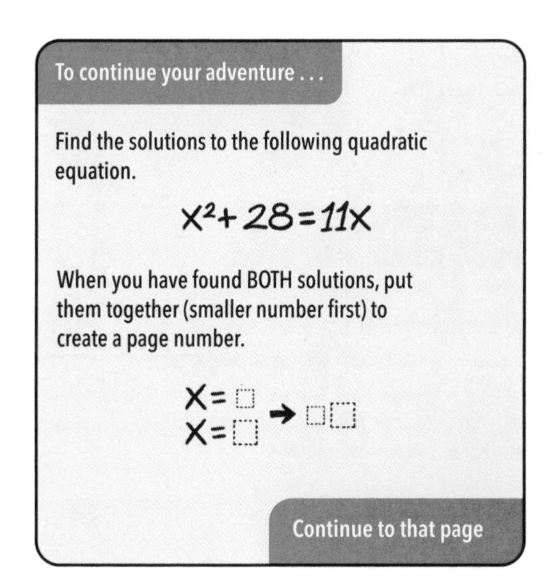

See page 58 of Adventurer's Advice Chapter 3 for help.

You should have arrived here from page 18 OR page 25

You climb up the side of the crane and quietly pull yourself into the cab. The seat is covered with long blue feathers. Who was operating this thing last? Below you, the robot is rounding the corner, and you can see a pointy light atop its head blinking orange. “Scanning . . . Scanning . . .” the robot repeats over and over in a tinny monotone. Its head turns in circles as it searches for the intruder. *As it searches for YOU.*

You stay low in the cab of the crane to avoid being detected, but how long can you wait before the robot finds you? And now you see below you two more robots have joined in the search. *You have to do something.*

You pull out your phone to text Hazar. Maybe he can help you get out of this mess. *No service*. Besides, you aren’t sure your cell phone plan covers inter-portal texting. You shove the phone back in your pocket, and as you do, you bump a few of the feathers from the seat out of the cab of the crane. They fall silently to the ground.

Below you, one of the robots spots a feather as it lands softly near the boxes of insulin. Without warning, the light on its head turns red, and a violent laser blast emits from the front of its torso. The feather is obliterated in a flash of red light, leaving behind a small crater in the cement. The robots seem dissatisfied with the feather, and the chorus of “Scanning . . . Scanning . . .” continues.

You know that if you wait for too long, the next laser blast is coming for you. You glance around the cab. To your surprise, the keys to the crane are in the ignition. It will make noise when you turn on the crane, but what choice do you have?

You turn the key in the ignition and the diesel engine of the crane rumbles to life.

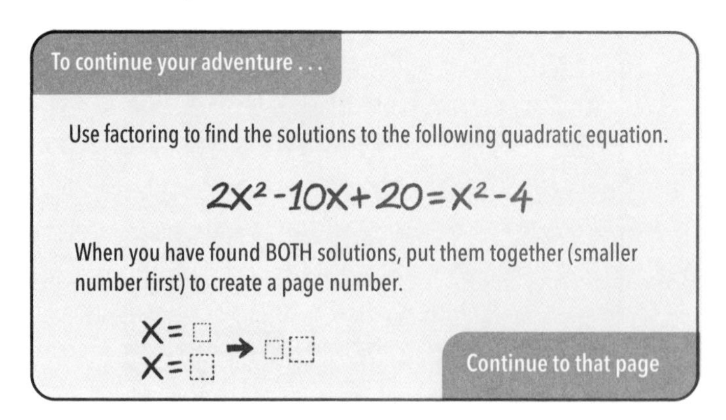

See page 58 of Adventurer’s Advice Chapter 3 for help.

You should have arrived here from page 11

This left hallway is carpeted with a plush, royal blue carpet that muffles the sound of your footsteps. You walk past a number of dark offices where the bankers and lenders must work during the day. A robot vacuum is quietly patrolling the offices and sucking up bits of dust and dirt.

You pass underneath a security camera and notice that the lens has been spray-painted over in bright pink paint. Whatever is going on here is *sketchy*, and you duck from doorway to doorway, in case you come across someone from the Dregg Corporation.

You pass a life-size cardboard cutout of a woman advertising the bank. She has dark hair and a smart blue blazer. The speech bubble coming from her mouth says, "Those who know, know to choose the bank where know-how knows no no-nos."

Wow. That's a terrible slogan, you think to yourself as you approach the server room.

To continue your adventure . . .

Solve the system below using substitution.

The parabola $y = x^2 - 8x + 18$

crosses the line $y = -x + 6$

in two places. One of those points is (3,3). Find the other point.

Turn that point into a page number, and continue to that page (☐,☐) → ☐☐

See page 65 of Adventurer's Advice Chapter 3 for help.

You should have arrived here from page 4

You feel terrible about the shock you just gave this bear, but you didn't do it on purpose. Besides, who put on the shock collar in the first place? What kind of heartless monster treats workers this way?

You turn your attention to the other screens in the room. There must be 60 screens. Each one shows a "Non-Human Resource" that is trapped in a shock collar, working for Dregg. Some are animals, like the bear. Many more are robots with spindly metal legs and pointy lights atop their heads.

Dregg has an army!

These latest developments are a little overwhelming, and they leave you with more questions than answers. What is with the boxes that the bear is moving? What is Dregg using all these animals and all these robots for?

You can't leave now. Not until you get some answers.

You turn back to the lab, and as you do, you notice a bright red spatula hanging from a nail by the door. You can see a key hanging from the hole at the end of the handle, and a piece of tape sticking off of the thing that has "lab" written on it in black marker.

You shake your head as you take the key and cross the hall to the lab. *The gas station by Taylor's house uses the same security system for their bathroom as Dregg does for this multimillion-dollar research facility.*

The key fits, and you unlock the door.

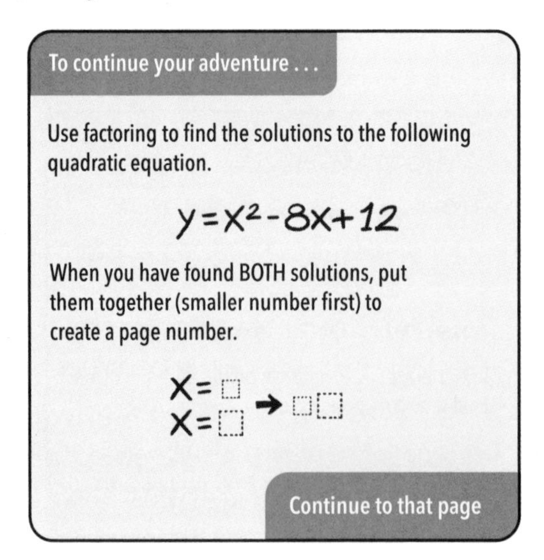

See page 56 of Adventurer's Advice Chapter 3 for help.

You move quietly among the boxes of insulin on the floor of the warehouse. The cardboard boxes are stacked well above your head and wrapped in cellophane. You hear a small clatter on the other side of the boxes stacked to your left. Two voices begin arguing in hushed tones.

You freeze and listen in to their discussion. ". . . and we can't repackage this stuff if it's broken! Don't be so clumsy! The insurance money doesn't matter if we have to spend it to make more insulin!"

Wait. *Really*? Dregg invented teleportation, and their goons are using it to steal their own prescription drugs. *Just to collect the insurance money*!

They *could* do so many great things with this technology.

Instead, they're doing petty crime.

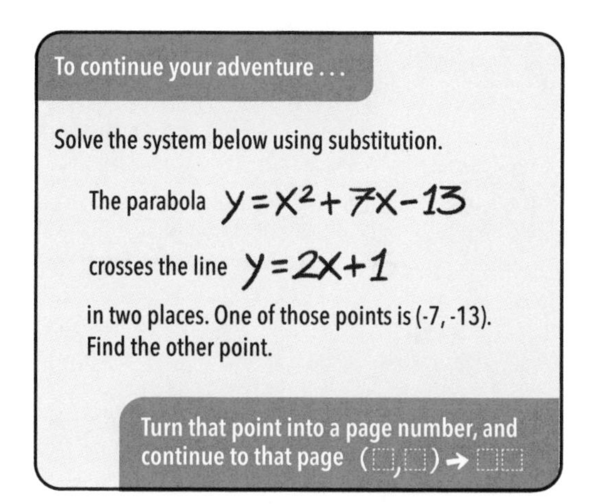

See page 65 of Adventurer's Advice Chapter 3 for help.

You should have arrived here from page 23

Think quiet thoughts. Think quiet thoughts.

As you crouch silently under the desk, the men are mere feet away searching the office. "Seriously, I swear I left my radio clipped to my pack."

You curse under your breath. *Why couldn't you have just left the radio clipped to the bag!?* One of the two men is checking the papers on top of the desk. They're going to find you at any moment. Just as you are about to give up hope, you can hear the phone in the office ring, and the man at the desk goes to answer it.

"Hello . . ." the man says into the phone. You eavesdrop on the half of the conversation that you can hear. ". . . #718 again? His name is Lazarus, right? . . . I don't care if he's a possum! . . . He can't play dead when we're in the middle of these things! He knows we're paying him for this job . . . *Yes*, I'll come and talk to him."

Both men exit the office, and you follow at a safe distance.

The warehouse is a little darker than the office, and you don't notice the robot moving boxes just down from you on the left. It spots you and shoots in your direction with a laser from its chest. It misses your torso by inches. You duck between two long rows of boxes, but you will need a better hiding spot soon. Just past the next row of boxes you can see a tall crane, and as you approach it you can see a shorter forklift sitting next to the crane. Either spot would be safer than here among the boxes.

You had better make a move quick. Before the robot finds you.

To hide in the crane, turn to page 14.

To jump on the forklift, turn to page 21.

Go back to the previous page and check your work. You should not have arrived here.

Go back to the previous page and check your work. You should not have arrived here.

CHAPTER 3

You should have arrived here from page 18 OR page 25

The forklift doesn't have doors, so this might not be the best hiding place after all. You climb in and duck low to hide yourself from view.

From your hiding spot, you watch the robot round the corner. Atop its head, a pointy light is flashing orange at two-second intervals. The robot's head slowly spins as it scans the area for an intruder. Ten feet from the robot, a fly lands on the ground, and the orange light switches to red. A powerful laser blast comes from a lens on the robot's torso, straight toward the fly. The blast leaves behind a smoking crater in the cement where the fly had been moments before.

The light atop the robot's head switches to a calm green color and you can hear "Threat terminated." You watch as the robot returns to its duties moving boxes across the warehouse.

That was too close. The next laser blast might be coming for you!

You have to get back through the portal.

You notice someone has left a Dregg Corp baseball cap on the seat, and also left the keys in the ignition. Maybe you could put the cap on and get to safety somewhere else in the warehouse? Or maybe you will be safer if you stay in the cab of the forklift. You could use the keys and try to turn this thing on.

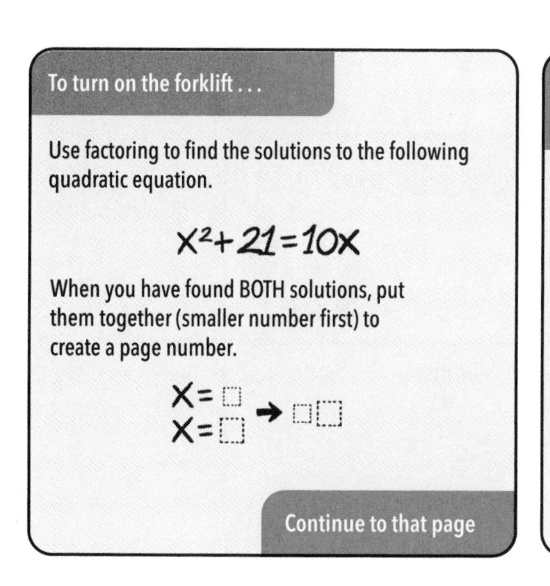

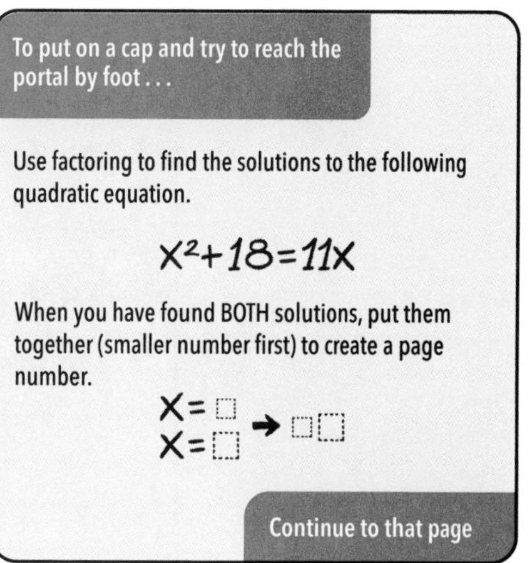

See page 58 of Adventurer's Advice Chapter 3 for help.

You should have arrived here from page 46

The robots that you accidentally picked up with the magnet are swinging back and forth wildly. All the robot legs are poking out at odd angles, making a tangled, spiky mess at the end of the crane. You press the red button with your thumb, and it stops blinking. The whole crane jolts as the magnet releases from the end of the cable on the crane.

The massive magnet and the hapless jumble of robots fall with a thunderous crash onto the stacks of boxes below. The sound of breaking glass echoes throughout the warehouse as dozens of boxes are crushed under the weight of the magnet. Dozens more rip open, showering the warehouse floor with thousands of tiny, expensive insulin vials.

The other Dregg thieves (two bears, one robot, and three humans) look on in horror at the scope of the destruction you have caused. When they look back to the cab of the crane, you are gone.

You run as fast as you can toward the portal. Its ethereal blue coils are still spinning and glowing at the far end of the warehouse.

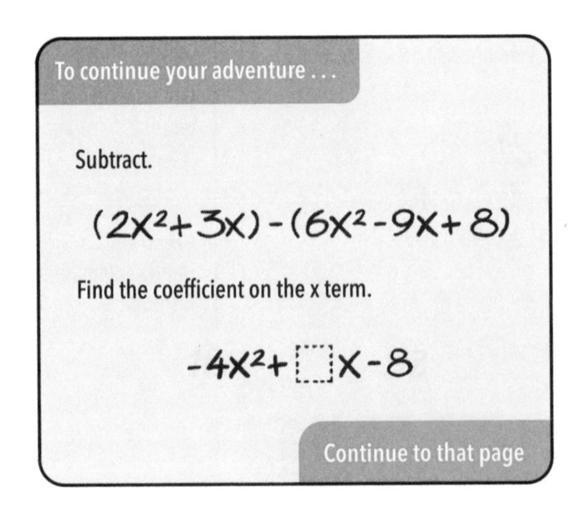

See page 53 of Adventurer's Advice Chapter 3 for help.

You should have arrived here from page 34

You grab the radio off the pack and slide under the desk just as the men begin climbing the stairs to the office.

They seem to be in a bad mood as they enter: ". . . and that better not affect my cut from tonight. I wasn't the one who made that mess out there."

"At least it wasn't Mico this time," says the other guy, and they both laugh.

"Hey, have you seen my radio?" asks a worker as he picks up his pack by the door.

Uh-oh. You glance at the radio in your hand. Your mind spins as you try to figure out what to do next. Maybe you could use the radio to distract them somehow and throw them off your scent? From under the desk, it doesn't seem like you have many other options.

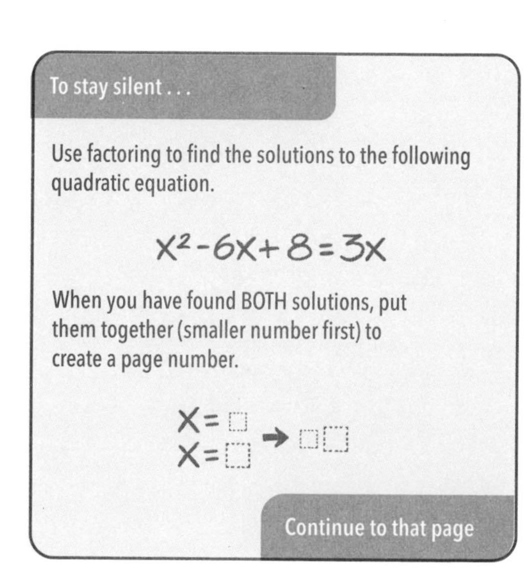

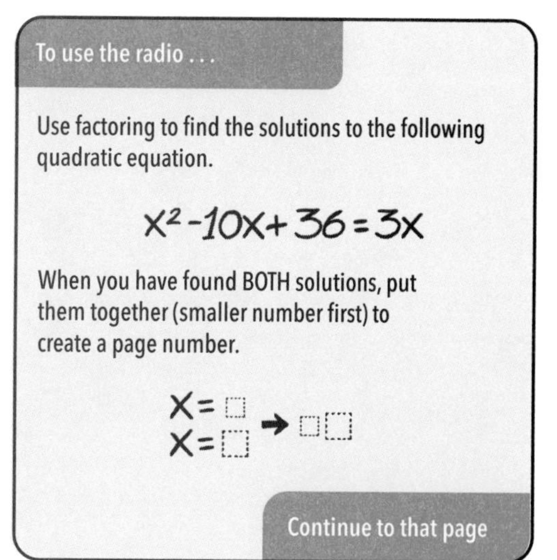

See page 58 of Adventurer's Advice Chapter 3 for help.

You should have arrived here from page 46

You grab the crane's steering wheel and spin it counterclockwise as hard as you can. The cab you're sitting in starts to spin, and with it so does the swarm of robots stuck to the magnet at the end of the crane. One of the robots fires a laser blast at you, but it misses badly. The magnet is moving too fast for them to properly aim.

The rest of Dregg's portal team is on the floor below the crane now. They have stopped what they are doing and are watching in disbelief as you joyride the crane. You have to admit, it's pretty fun.

The crane swoops in a wide arc and speeds through the air as you continue to spin. As you near a full rotation, the magnet and the robots crash violently into the side of the elevated office structure. The collision crushes the robots and sends a shower of splintered wood and broken glass onto the warehouse floor below.

You let go of the steering wheel and it spins clockwise, back toward its starting position. The cab of the crane and the magnet spin with it, flailing wildly in wide, chaotic arcs through the warehouse.

Dregg's goons watch you, speechless. No one can stop what you've started. The magnet knocks over boxes of insulin, knocks out overhead lights, and carves a huge gash in the nearest wall. At this point, you'll be lucky if you survive to answer for the damage you've created in the Dregg warehouse. Looks like you've reached . . .

The End

You should have arrived here from page 17 OR page 43

You continue across the warehouse floor and glance back toward the portal, where a couple of spindly robots are pushing boxes of insulin stacked on dollies toward the portal. The robots look just like the ones that you saw in Dregg Tower, with pointy lights atop their heads and cameras atop six spindly legs.

One of the robots turns its head to face you, and your heart drops. *Have you been spotted?* You crouch down behind a tall stack of boxes, and you can hear the the robot walking toward you to investigate.

click*clack*click*clack*click*clack

Its pointy metal legs make loud skittering noises against the hard cement floor, and they are growing louder as the robot approaches.

You need a distraction, or at least a better place to hide. Beside the long rows of boxes, you notice a few pieces of heavy machinery here on the warehouse floor. There is a gigantic crane that they must use to move heavier stuff. You also see a forklift. It's just past the crane.

You had better make a move quick. Before the robot finds you.

To hide in the crane . . .

Find the vertex of the following parabola.

$$y = -2x^2 + 4x + 2$$

Turn your solution into a page number, and continue to that page (☐,☐) → ☐☐

To hide in the forklift . . .

Find the vertex of the following parabola.

$$y = x^2 - 4x + 5$$

Turn your solution into a page number, and continue to that page (☐,☐) → ☐☐

See page 63 of Adventurer's Advice Chapter 3 for help.

You should have arrived here from page 16

With the doors open, you pass into a huge lab space with high vaulted ceilings. You call out, but there doesn't seem to be anybody in here. And what a mess! Giant cardboard boxes, piles of paper, empty cans and mugs, and stacks of paintings leaned up against the wall. *How could anyone possibly get any work done in here?*

The lights are off, but there is bright blue light coming from the center of the lab. Three enormous metal structures have been assembled there, each with a ten-foot circular opening in the middle.

The first opening is empty, and you can see a chalkboard behind it where someone has scrawled lines and lines of complicated mathematical equations in messy shorthand. The center structure and the one on the right are each filled with eerie, pulsing circles of blue plasma.

CHAPTER 3 GOAL: Follow the blue plasma and learn what's REALLY happening at Dregg Tower.

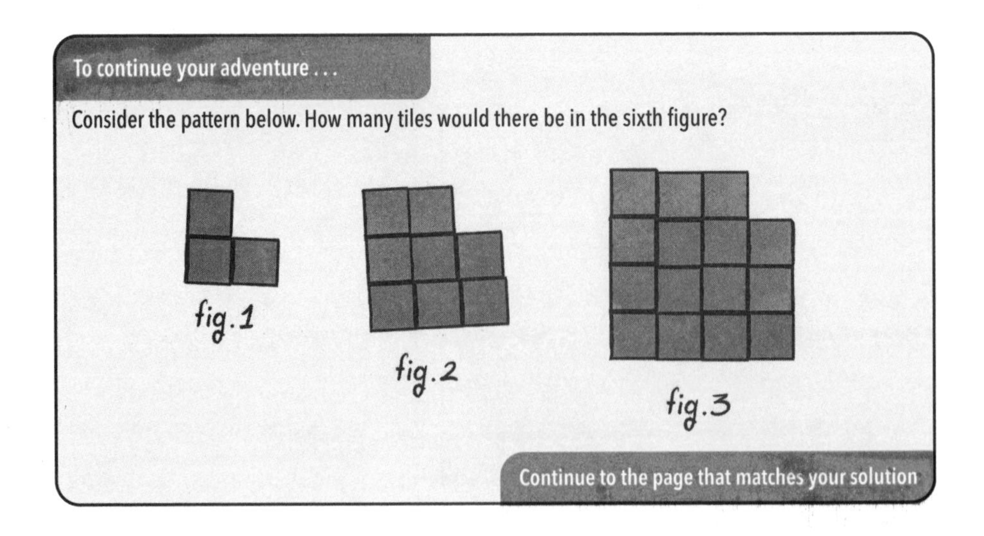

See page 55 of Adventurer's Advice Chapter 3 for help.

With the server room now empty, you sneak in to get a closer look at what Dregg was doing.

There is a laptop computer sitting on a table in the back of the server room, and it is connected to one of the black server boxes by a colorful array of wires.

On the laptop, a program is open and running. A window reads “Download 80% complete.” Behind the window, you can see a list of names rapidly scrolling by. *Waldron . . . Wapato . . . Waverly . . .* It’s in the W’s right now. You scroll down and realize quickly this is a data transfer from the bank’s servers.

These Dregg henchmen are downloading the banking information for every member of the bank! There is way more money in those personal accounts than there is in the vault. What a devastating turn of events for the people on this list. Does Hazar’s mom use this bank, or are you remembering it wrong?

There is a button on the screen that says “cancel download” and one right next to it that says “eject.” If you cancel the download, you know the members of the bank won’t lose their savings. But . . . if the other button ejects a hard drive, you will have all the evidence you need when you finally get out of this bank.

To click “cancel download” . . .

Use the discriminant to find the number of real solutions to this equation.

$$y=11x^2-10x+7$$

If it has ***0 solutions***, continue to page 5
If it has ***1 solution***, continue to page 20
If it has ***2 solutions***, continue to page 3

To click “eject” . . .

Use the discriminant to find the number of real solutions to this equation.

$$y=-7x^2-7x+10$$

If it has ***0 solutions***, continue to page 35
If it has ***1 solution***, continue to page 19
If it has ***2 solutions***, continue to page 44

You should have arrived here from page 33

You put your shoulder against the heavy vault door and push with all your weight. It slowly begins to move on its hinges, and slams shut with a loud click. There are a few spinning handles on the front of the door. You don't know what they do, but you spin them for good measure.

Does righty-tighty work with bank vault doors?

You can hear the monkey bank robbers shriek and pound on the door as they realize that they are locked inside.

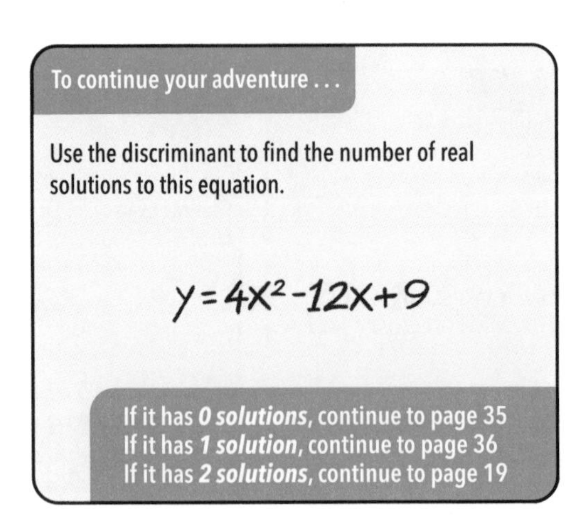

To continue your adventure . . .

Use the discriminant to find the number of real solutions to this equation.

$$y = 4x^2 - 12x + 9$$

If it has ***0 solutions***, continue to page 35
If it has ***1 solution***, continue to page 36
If it has ***2 solutions***, continue to page 19

See page 61 of Adventurer's Advice Chapter 3 for help.

The key to the forklift has a little toy race car attached by the key ring. *This is the energy you need.* You turn the key and coax the engine to life. It spits out big plumes of dark black smoke from the exhaust and it makes a loud, shrill beeping noise as you put it in reverse. With the beeping and the smoke, it's a minor miracle you aren't spotted right away.

You turn the forklift toward the portal and push the accelerator to the floor. The tiny wheels hum along the smooth cement floor as you speed off. At this rate, you'll be back at Dregg Tower in no time!

Up ahead, you see two of the Dregg robots pushing boxes, and they spot you immediately. They turn from their work and block your pathway to the portal. The first one charges up a laser blast, and you realize that the cab of the forklift won't provide you much cover. You can't slow down or you're going to be an easy target!

You could continue straight toward the robots, and if you're lucky you'll be able to bully your way past them.

You could also swerve out of the way. At this speed, you could probably crash through the cardboard boxes to your left. But what's on the other side?

To continue toward the robots . . .

Use the discriminant to find the number of real solutions to this equation.

$$y=2x^2-4x+2$$

If it has ***0 solutions***, continue to page 35
If it has ***1 solution***, continue to page 40
If it has ***2 solutions***, continue to page 19

To swerve toward the cardboard boxes . . .

Use the discriminant to find the number of real solutions to this equation.

$$y=6x^2-3x+3$$

If it has ***0 solutions***, continue to page 2
If it has ***1 solution***, continue to page 20
If it has ***2 solutions***, continue to page 3

See page 61 of Adventurer's Advice Chapter 3 for help.

You should have arrived here from page 49

A peculiar group is gathered by the portal, furiously discussing something. There are three giant grizzly bears, flanked by four of the spindly robots. Three human workers are there too, yelling back and forth with a monkey. By now they must have discovered that there was no second spill, and that there is an unwanted visitor somewhere in the warehouse with them. They are looking around uneasily.

You know that if they leave through the portal you will be stranded here, so you grab the radio again. "Come quick! There's a little *punk* up here in the office!"

The strange group of workers scurry toward the office to investigate. *Dregg needs better henchmen on its Secret Portal team*, you think. *That shouldn't have worked twice.*

This is your chance. You drop the radio and dash toward the open portal at the end of the warehouse.

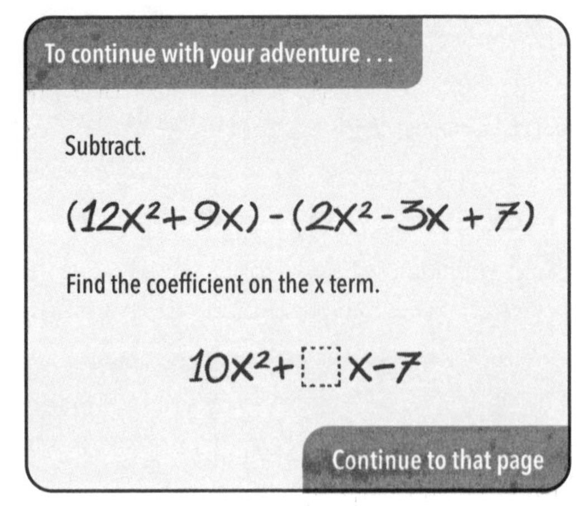

See page 53 of Adventurer's Advice Chapter 3 for help.

You should have arrived here from page 42

Lure them out, *but how?* Your mind races trying to figure out a distraction. You dart back down the hall and collect the robot vacuum and the life-size cardboard cutout of the woman with the confusing corporate slogan. You quickly glance through the drawers of one of the desks and are able to find clear tape and a book of matches.

Using some tape, you attach the cardboard woman to the top of the robot vacuum, and she begins to glide around the hallway. Each time she collides with the wall, the robot vacuum reroutes in a new direction. You point her toward the server room and light a match. You hold the match to the dry cardboard cutout, and it begins to burn and smoke while awkwardly shuffling about the hall on the robot vacuum. You hide in a doorway before bellowing out in your most sinister voice, "*Intruders! You should have had the know-how to know that this is a no-no!*"

The mysterious figures in the server room emerge to a smoking, flaming menace gliding toward them. In the light of the hallway you can see exactly who Dregg had sent to raid the server room: two llamas, rigged up with strange little backpacks, silver helmets, and shock collars. They bolt down the hall, bleating in fear.

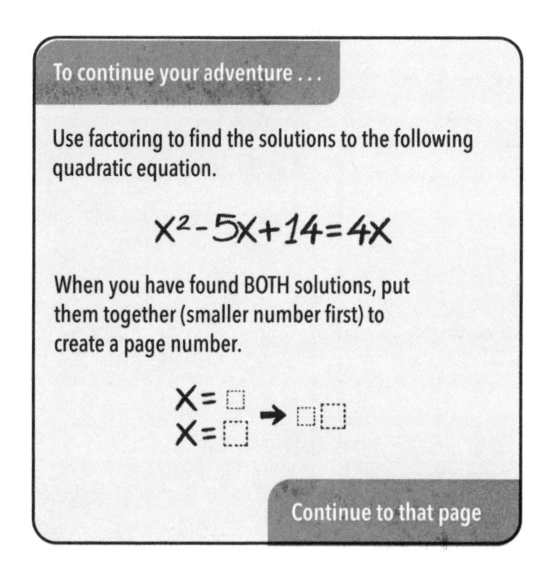

See page 58 of Adventurer's Advice Chapter 3 for help.

You should have arrived here from page 6

You step softly across the purple and teal tile to the metal gate that leads to the bank's enormous vault. Someone has used a welding torch to melt through the bars on an ornate security gate, and past the security gate you can see the tall, circular vault door.

The vault door is slightly ajar, and you can hear what sounds like a powerful vacuum coming from inside. You creep through the melted bars of the metal gate to inspect a small pile of equipment next to the vault door. There is a stethoscope, a radio, a can of pink spray paint, and a black duffel bag with the Dregg logo emblazoned on the side. *Really seems like they could have found a different bag.*

You could see what's inside the bag, or you could get a closer look in the vault.

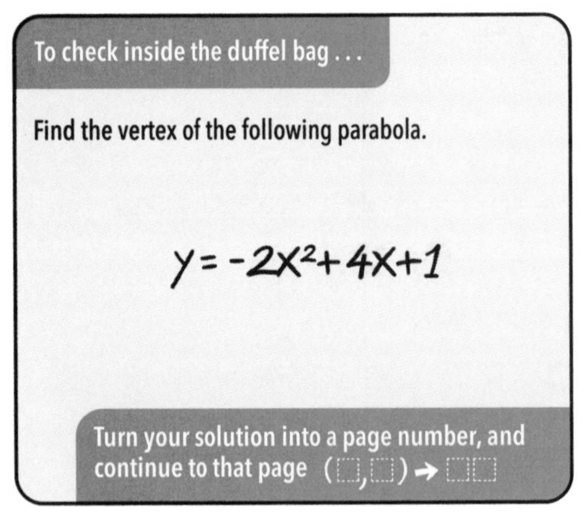

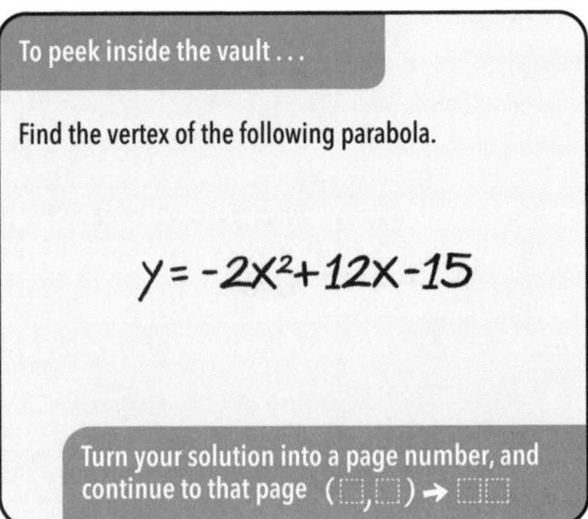

See page 63 of Adventurer's Advice Chapter 3 for help.

You should have arrived here from page 32

The huge round door is cool to the touch, and you carefully peer around its edge into the bank vault. It is a twenty by thirty foot room, with safe deposit boxes lining the walls and piles of cash in the back that the bank is keeping on hand. It's the bank robbers, though, that you notice first.

Seven furry little capuchin monkeys are racing around the vault, while two hulking gorillas stand in the middle of the room.

One of the capuchin monkeys is using a huge ring of keys to open the safe deposit boxes.Three others are climbing up and down the racks of boxes and dumping their contents onto the floor. The rest of the capuchin monkeys are sorting the valuables into burlap sacks. The gorillas are carrying giant vacuum cleaner packs on their backs. Each vacuum pack has a long hose that the pair of primates are using to suck up money from the neatly stacked piles of cash along the back wall.

Should you slam the vault shut, locking this monkey business inside? Or just make a run for it, before you get further involved?

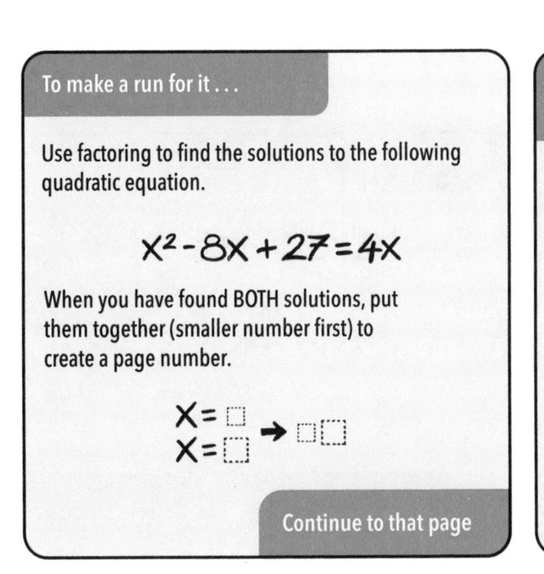

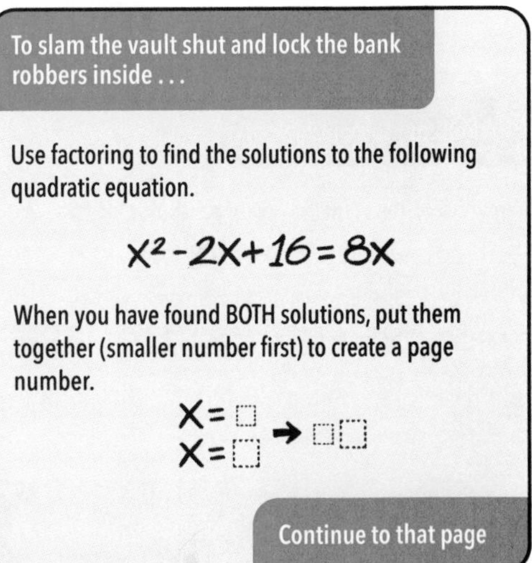

See page 58 of Adventurer's Advice Chapter 3 for help.

You should have arrived here from page 7

You spin around. *Who said that?* You are alone in the office. There is a backpack near the door that you hadn't noticed before. Hanging off of the pack is the source of the voice, a small radio. The thieves must be using them to communicate.

"You gotta be more careful! We can't repackage this stuff if it's broken! Don't be so clumsy! The insurance money won't matter if we have to use it to make more insulin!"

Wait. *Really*? Dregg invented teleportation, and their goons are using it to steal their own prescription drugs. *Just to collect the insurance money!* They could do so many great things with this technology, but they're using the portals for petty crime.

"Get your gear and let's get out of here," says the voice over the radio.

You look out the window to see two men coming right for you, and the door is no longer an option. You could climb out the window to avoid detection, or grab the radio and hide under one of the desks.

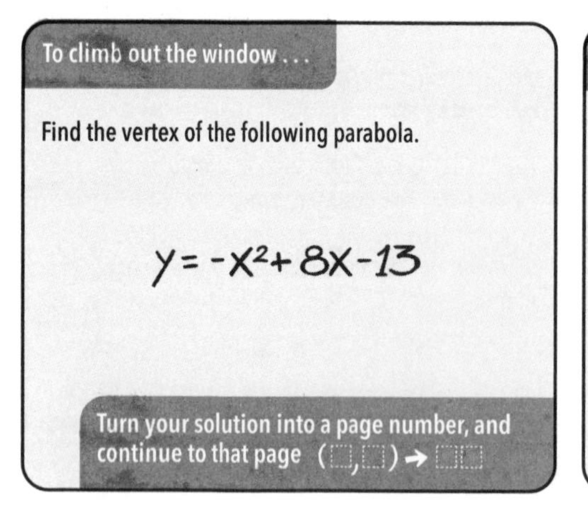

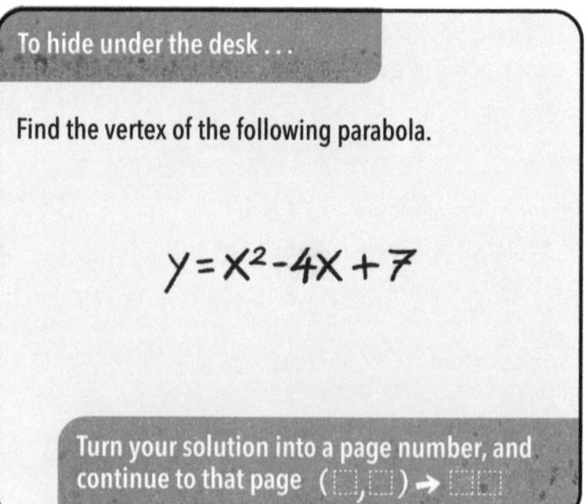

See page 63 of Adventurer's Advice Chapter 3 for help.

Go back to the last page and check your work. You should not have arrived here.

You should have arrived here from page 28 OR page 47

The bank robbers are pounding loudly on the vault door, and you know that if there are any other Dregg goons nearby, they will hear the pounding soon.

Who knows, they may even be able to call for help with another radio. *Time to get moving.*

You make your way back down the broad corridor and into the bank lobby.

As you reach the lobby, you hear voices and loud footsteps in the corridor behind you.

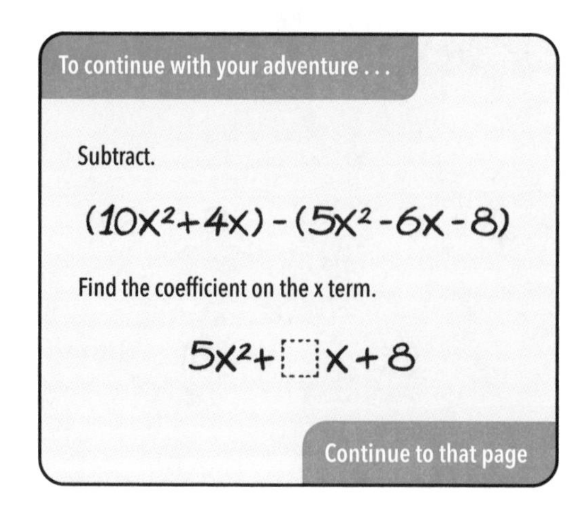

You should have arrived here from page 21

You pull the cap down low over your eyes and check your reflection in the forklift's rearview mirror. *The mirror!* If you run into any more of those robots, it might be able to redirect a laser blast.

The adhesive holding the mirror to the windshield is old, and you are able to pull the mirror down without too much trouble. You feel safer with the mirror, but does that actually work anyway? You've seen it work in movies, but that robot you saw earlier was very much *not* from a movie.

You drop from the forklift to the warehouse floor and sneak between the rows of boxes toward the portal, gripping the mirror in your left hand.

There is a break in the boxes, and you stop to peer around the corner. *Make sure the coast is clear.* You hear a noise from behind you.

"Threat identified."

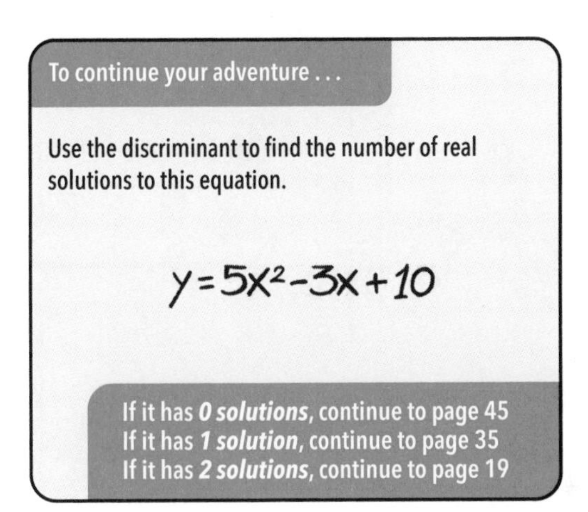

See page 61 of Adventurer's Advice Chapter 3 for help.

You chance a peek from your hiding spot toward the door into the hall. The workers are not at all what you had been expecting. Dregg has two llamas working this server room. Each one is wearing a silver helmet with colorful wires running down their long, fuzzy necks. A black backpack is strapped to each llama's back. One of them spits on the floor near your feet. *Gross.*

They walk past you and into the hall, searching for the source of the flute music, but your phone is safely back in your pocket. *This might be the only chance you get!* You dart over to the computer where they had been standing and look at the screen. "Download 55% complete." Behind the window, you can see a list of names rapidly scrolling by. *Paul . . . Pauli . . . Pauling . . .* It's in the P's. You scroll down and realize quickly that this is a data transfer from the bank's servers.

These *llawless llamas* are downloading the banking information for every member of the bank! These personal accounts have far more money than the vault down the hall! What a devastating turn of events for the people on this list.

There is a button on the screen that says "eject" and one right next to it that says "cancel download." If you cancel the download, you know the members of the bank won't lose their savings. *But . . .* if the other button ejects a hard drive, you will have enough evidence to bring down Dregg when you finally get out of this bank.

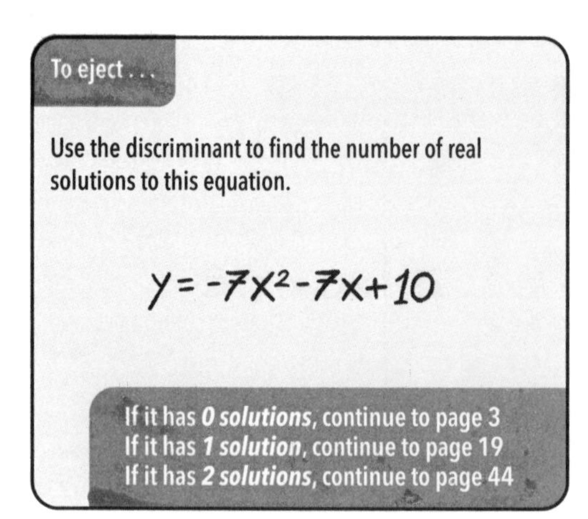

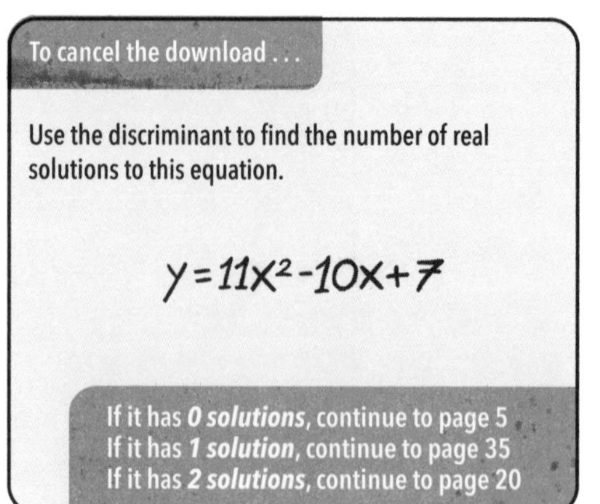

See page 61 of Adventurer's Advice Chapter 3 for help.

You should have arrived here from page 33

You make a run for it! Who knows how long the portal will stay open, and you are completely outnumbered.

You hear shouting and rustling as the bank robbers spot you and chase after you. The capuchin monkeys aren't as much of a concern, but each of the gorillas outweighs you by at least 200 pounds. They are chasing you now too, and one of them is brandishing the tube to the vacuum pack like a sword.

Panicked, you look around you. There's a restroom door to your left—maybe you should dive in to escape. No way you can make it all the way to the lobby.

To dive into the restroom, turn to page 50.

You should have arrived here from page 29

You head straight for the two robots standing between you and the portal. One fires a laser blast at you and it barely misses, melting a baseball-sized hole in the seat of the forklift just to your right.

You raise the fork sticking off the front of your machine so that it is level with the robots. *You're only going to get one shot at this.*

The gas pedal for the forklift is already against the floor panel of the machine, but you push down harder with your foot anyway. *This has to be perfect!*

One of the robots is charging up another laser blast, and you catch him through the torso with the left fork of your lift. The robot is impaled on your fork, and the laser blast is due any second.

Calmly, you channel all of your skills from driver's ed. *Hands at ten and two. Check your mirrors. Take a deep breath.* You execute a perfect three-point turn so that the robot on the front of your machine is pointing toward the other robot *just* as the laser fires. It melts a perfect hole through the head of the second robot, causing it to collapse in a pile on the floor.

Two robots down! You parallel-park the forklift between two rows of cardboard boxes and make a beeline for the portal.

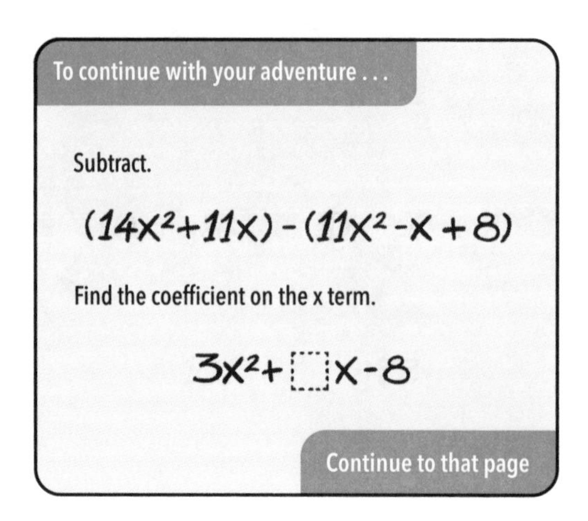

See page 53 of Adventurer's Advice Chapter 3 for help.

You should have arrived here from page 42

You take out your phone and look at it for a moment.

Taylor nudged you again on Slime Ninja. She must be by her phone. You fire off a quick text.

Taylor, I need a favor.

A moment later, her response pops up.

Ugh. Take your turn and THEN we can talk.

She's stubborn. You know that there's no way you will get any help until you go. You open the Slime Ninja app, and the familiar pan-flute music starts playing. There are intermittent slashing sounds as a sword cuts back and forth across the screen. Your volume is turned all the way up, and it's loud. Panicked, you turn off the volume, but it's too late. The two Dregg workers in the server room heard you for sure.

There is no cover here in the hall, and you'll have to move quickly. As they turn toward you, you are able to scramble inside the server room and duck behind the last row of servers. It's dark, and you are pretty well hidden between the server and the wall.

It won't matter how good your hiding spot is if they saw you, though. Have you been spotted? Is this the end?

Shhhhhhhhh . . .

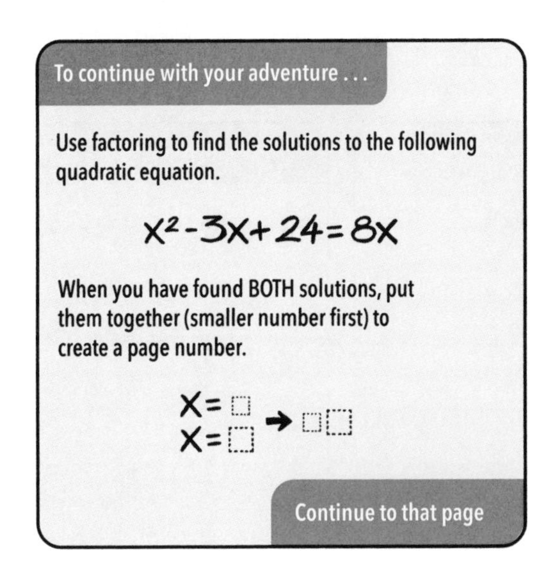

See page 58 of Adventurer's Advice Chapter 3 for help.

You should have arrived here from page 15

You continue down the hallway toward the server room. The door is cracked open a bit, and there is light spilling into the hallway.

You peek inside and can see rows and rows of tall black computer servers. There are mazes of wires connecting the tall black boxes, and small blue LEDs blink on and off as the machines quietly whir.

You see two strange figures in the back of the room. Their backs are turned to you and it's dark—you can't make out their features.

You pause outside the room. Whatever happens next, you will be outnumbered. Your only advantage is the element of surprise.

Maybe you could come up with a way to lure them out? You're pretty clever. Or maybe you could get some help from one of your friends? Maybe you get phone service in here.

To lure the intruders out of the server room . . .

Find the vertex of the following parabola.

$$y = 2x^2 - 12x + 19$$

Turn your solution into a page number, and continue to that page (□,□) → □□

To try to message one of your friends . . .

Find the vertex of the following parabola.

$$y = 2x^2 - 16x + 33$$

Turn your solution into a page number, and continue to that page (□,□) → □□

See page 63 of Adventurer's Advice Chapter 3 for help.

You should have arrived here from page 34

You time your escape out the window carefully. You wait until the two men coming to the office reach the stairs. Their footsteps on the creaky steps drown out the sounds of you climbing out the window and dropping to the warehouse floor below. You sneak low across the warehouse floor and begin to snake your way between the neatly stacked boxes back toward the portal. You take a right, then a left, then another right.

The last corner you take a little too quickly, and you run headfirst into something warm and soft. The collision knocks you back. *Oh no.* It's a bear like the one that you saw in the security camera footage!

Actually, the number on the helmet is 517, and it has the same white streak in its fur. *This is that bear.* The bear takes a step toward you and you try to back away, but you are up against a wall of cardboard boxes. The bear has you cornered. It steps toward you and growls angrily. You close your eyes. *This might be it . . .* The bear pauses and glances up at the camera on the wall. Then it grumbles and returns to loading boxes onto a dolly. *It doesn't want another jolt to the shock collar.*

Dregg's inhumane methods with this poor bear just might lead to their downfall. *But you have to get out of here to tell your story first.*

To continue your adventure, turn to page 25.

You should have arrived here from page 27 OR page 38

You search the sides of the computer for the USB thumb drive. It *must* be here somewhere, but all you find are a few thick red wires and an old floppy-disk drive from the 1980s. *This bank needs to update their system,* you think to yourself. Maybe a disk is what "ejects" when you click the button. No problem. That will work as evidence just as well as a thumb drive.

Only one way to find out.

You click the "eject" button on the screen and nothing happens.

You click it again, and this time it takes. There is a loud bang below the floor and then a rush of air past your face as you are launched out of the bank and hundreds of feet up in the air.

I'm the thing that ejects!? You wonder. *Why did that server room have an eject button like an airplane?*

You fly high out of the bank and there is nothing you can do to prevent your fate. Gravity brings you to . . .

The End

You should have arrived here from page 37

You spin around to see one of the tall robots standing fifteen feet behind you. How did he creep up with those pointy metal feet on this cement floor? You turn to run, but there is another robot between you and the portal. Two robots, one mirror.

Even with your baseball-hat disguise, they have you identified as a threat, and each one is charging up a laser blast.

You look back and forth between them, gripping the mirror tightly. *Hopefully this works.* Which one do you even point the mirror at? Either way, your back is unprotected!

At the last second you dive out of the crossfire as the two robots shoot. One of the robots catches a laser blast to the chest and crumples to the floor. The other loses two legs and collapses on its side. It is still flashing red from its pointy head and chirping at you. "Threat detected! Threat detected!"

Forget the mirror—if you run fast, you should be able to make it to the portal before this robot can get another shot.

With the robots smoldering on the floor, you run as fast as you can across the warehouse, toward the spinning portal, and back toward Dregg Tower.

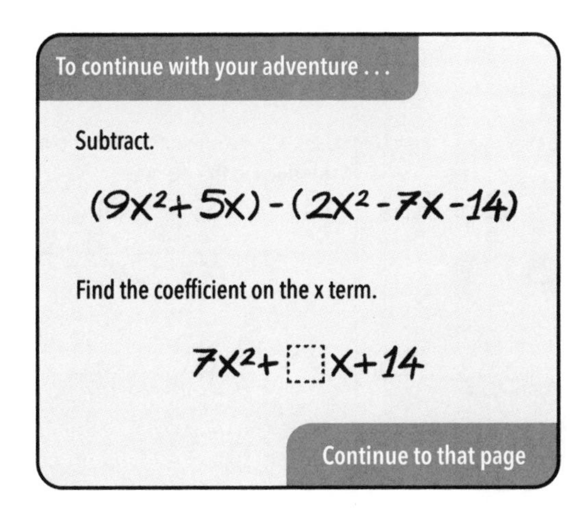

See page 53 of Adventurer's Advice Chapter 3 for help.

You should have arrived here from page 14

The crane is operated by long handles with round knobs on the end. The controls are jerky at first, and the large, flat attachment on the end of the crane swings about dangerously.

The robots below are freaking out. All three are flashing red light from their pointy heads and sporadically blasting the swinging end of the crane with their lasers.

The attachment on the end of the crane swings back and forth at the end of its cable, caught in the crossfire. It swings left, and as it comes back to the right, it crashes into the three robots below you. As you pull back on the controls, all three robots are lifted high in the air, flailing their spindly metal legs wildly. *The crane must have a strong magnet at the end!* And look at what you caught!

You turn your attention back to the controls. There is a big red button that is blinking. There is also a steering wheel that you haven't used yet.

To push the red button . . .

Use the discriminant to find the number of real solutions to this equation.

$$y = 6x^2 - 3x + 3$$

If it has ***0 solutions***, continue to page 22
If it has ***1 solution***, continue to page 3
If it has ***2 solutions***, continue to page 35

To spin the steering wheel . . .

Use the discriminant to find the number of real solutions to this equation.

$$y = -3x^2 - 8x + 12$$

If it has ***0 solutions***, continue to page 19
If it has ***1 solution***, continue to page 20
If it has ***2 solutions***, continue to page 24

See page 61 of Adventurer's Advice Chapter 3 for help.

From your vantage in the darkened office, you can see three figures, shadowy ones, approach the vault and creep inside. *Does that last one have a tail?*

You creep across the hall just in time to see the three figures pass by the big round door and into the vault. *That last one definitely has a tail.*

You spring into action, throwing your weight against the heavy metal door of the vault. It swings shut slowly and makes a satisfying "clank" sound, locking the bank robbers—and Beta unit—inside.

There are some dials on the front of the vault door, and you spin them for good measure.

If it wasn't locked before, it sure is now.

You can hear pounding on the inside of the vault door as the robbers realize they have been locked inside.

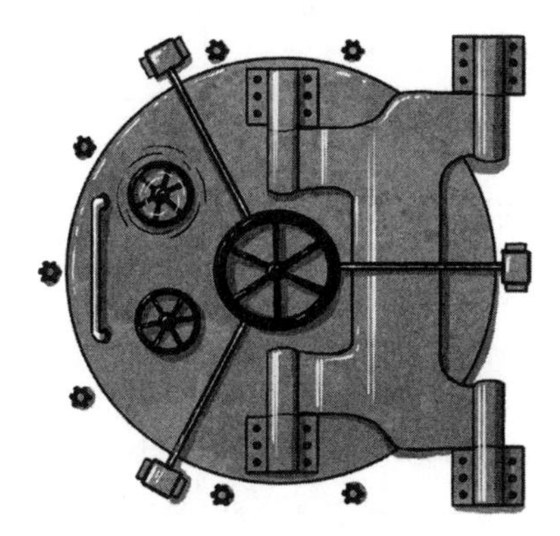

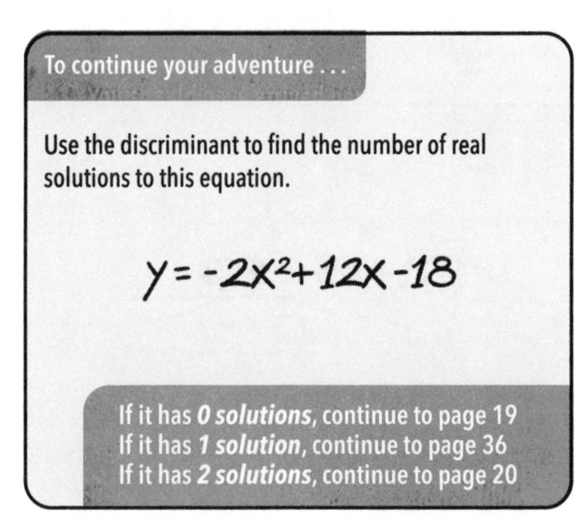

See page 61 of Adventurer's Advice Chapter 3 for help.

You should have arrived here from page 26

As you pass the nearest lab desk, you notice a ledger with the name "LaBella" embossed in the top corner. *Her desk! You made it!* You peek at the first page of the bound notebook and read a short note scrawled in red ink: Dregg has taken my research, and I don't trust what they are using it for. I may be in danger. -DL

Suddenly, you get a text. Hazar! You almost forgot you had asked for his help.

Not too much on Doctor LaBella, Hazar texts. First in her class at Larney University. PhD in Robotics. She has worked for Levi Dregg for a long time. Seems pretty loyal. She interned at Dregg Corp and she has been on the research and development team for years.

Degree in robotics? There were a bunch of robots in that security camera footage. Whatever is going on here, Doctor LaBella is tangled up in it. You still want to return the wallet for the reward money, but after everything you have seen, can you trust her?

You take a step backward to think, and as you do, you trip over a burlap sack on the floor. Your foot stings a bit. Something inside is heavy. You pull the bag open, and to your disbelief it is filled with large gold bars. Probably twenty of them. You look around the room and can make out dozens of similar burlap bags stacked against the walls and littering the floor near the blackboard.

Where did all this come from? If all these bags are filled with gold, they must be worth millions of dollars! You check another bag, and this one is filled with jewelry. Necklaces, rings, loose diamonds. What is going on here? Your attention shifts to the swirling blue light coming from the center of the lab. You walk closer to investigate.

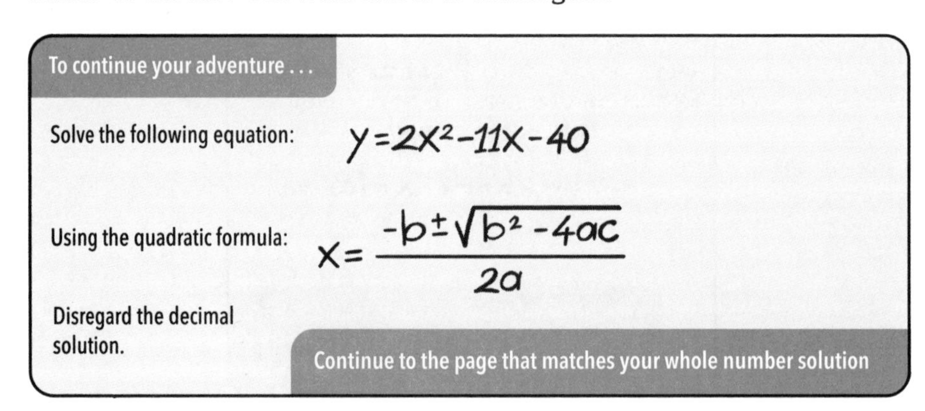

To continue your adventure . . .

Solve the following equation: $y=2x^2-11x-40$

Using the quadratic formula: $x=\frac{-b\pm\sqrt{b^2-4ac}}{2a}$

Disregard the decimal solution.

Continue to the page that matches your whole number solution

See page 60 of Adventurer's Advice Chapter 3 for help.

You should have arrived here from page 23

You push the button on the side of the radio to talk and try out your gruffest angry voice.

"Mico! Not again! Everybody to the portal. He spilled another box," you say into the radio.

"Gahhhh! That baboon buffoon!" The two men in the office exit out the door cursing, and you follow at a distance so as not to be discovered. You keep the radio in one hand.

What are they going to do when they discover your little lie?

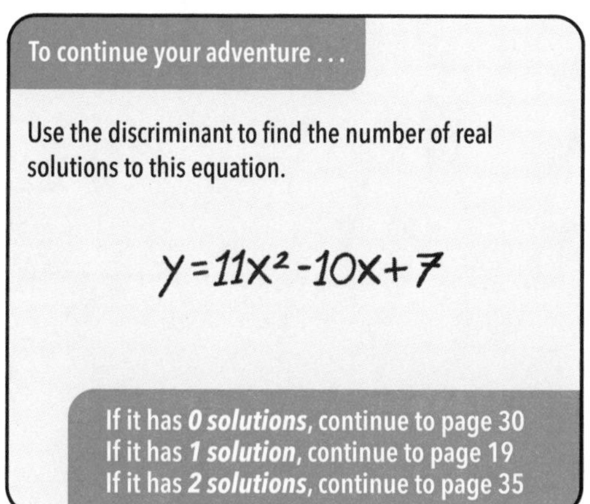

See page 61 of Adventurer's Advice Chapter 3 for help.

You should have arrived here from page 39

You fling open the restroom door, leap inside, and throw your whole body against it while the monkeys pound to get in. The door jolts as one of the gorillas smashes his hand against the door, but you are able to get the dead bolt on the door locked before they can pry the door open.

Stupid monkeys.

After a few minutes the pounding on the door stops altogether. *Maybe the monkeys gave up.* You pause with your ear to the door and listen intently. Is the coast clear? You can't hear anything outside the door, but you do notice a scratching sound coming from the ceiling, and it's getting louder.

All of a sudden, the ventilation grate in the ceiling pops loose, and the seven capuchin monkeys from the vault drop to the marbled floor of the bank bathroom one at a time. They snarl and show off pointy little rows of teeth. *You should not have underestimated the capuchins!*

One of the little monkeys darts past you and unlocks the dead bolt, letting the two massive gorillas into the bathroom. Their eyes are angry, and this might be the end for you.

Stupid monkeys.

The End

Quadratic vocabulary (page 1)

In which *quadrant* is the *vertex* of this *parabola* located?

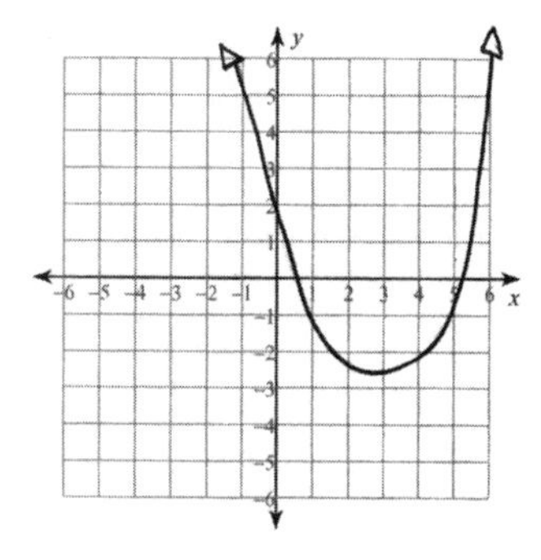

STEP ONE: This question uses a few pieces of key vocabulary that we need to understand if we are going to find the answer. First off: a parabola. Any function or pattern with x^2 as the biggest term is called "quadratic" and the shape that it makes on a graph is called a parabola. Some parabolas are fat, some are skinny, but they are always shaped like a "u." This chapter focuses on the math that we can use with parabolas and quadratic functions.

STEP TWO: This u-shaped parabola can open upward or downward, but either way, we will get one point that is lower (or higher) than all the other points on the parabola. This point is called the vertex.

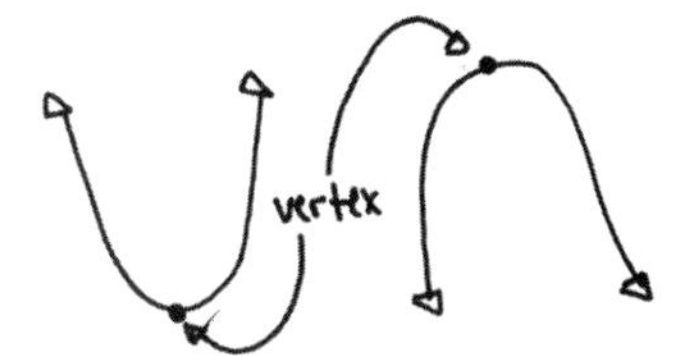

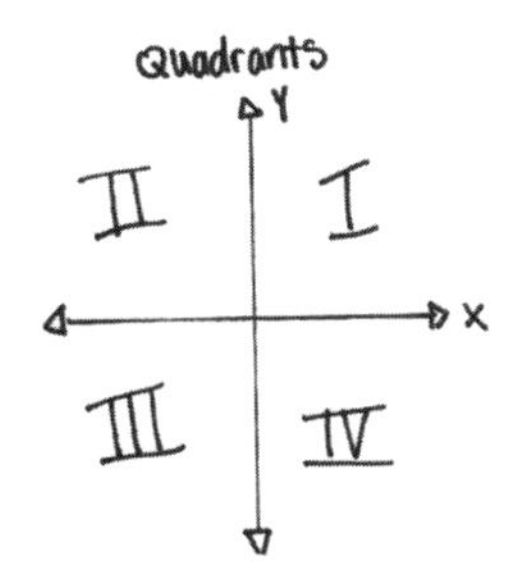

STEP THREE: Finally, we need to think back to what we know about a coordinate plane. The x and y axes split the coordinate plane into four parts called quadrants, and we typically number them 1–4, like the labels shown at left.

When we put all those things together, we can see that the vertex of the parabola shown in this question is in quadrant 4, so we will continue to page 4!

ANSWER KEY FOR THIS TYPE OF PROBLEM:

Page 1: Continue to page 4

Polynomial addition (page 4)

This is an addition problem, even though it looks more confusing. We need to add one thing $(7x + 1)$ to another thing $(4x^2 + 9x - 5)$. These things are called polynomials, and every polynomial has terms that are separated by addition or subtraction. So $(7x + 1)$ has two terms, $7x$ and 1.

$$(7x+1)+(4x^2+9x-5)$$
$$4x^2 + \square x - 4$$

$$(7x+1)+(4x^2+9x-5)$$
$$7x+1+4x^2+9x-5$$

STEP ONE: Add. When we add the two polynomials together, we need to add the $4x^2 + 9x - 5$ terms from the first polynomial to the $7x + 1$ terms from the second polynomial.

STEP TWO: Combine like terms. We have a mess of terms now and we can make our answer simpler by combining like terms. Like terms need to have the same variable *and* the same exponent. Otherwise, we can't count them together. In our problem, $4x^2$ doesn't have any like terms, but $9x$ does. The first polynomial has a $7x$ that we can combine with $9x$ for 16 total x's. -5 can also be combined with $+1$, so our final answer will be $4x^2 + 16x - 4$. Our page number is the coefficient with the x, so we will continue to page 16!

$$7x+1+4x^2+9x-5$$
$$16x+1+4x^2-5$$
$$16x-4+4x^2$$
Rearrange
$$4x^2+16x-4$$

ANSWER KEY FOR THIS TYPE OF PROBLEM:

Page 4: Continue to page 16

Polynomial subtraction (pages 2, 5, 22, 30, 36, 40, and 45)

This is a subtraction problem, even though it looks much more complicated. We are starting with one thing ($9x^2 + 5x$) and subtracting away another thing ($4x^2 - 7x - 5$). The "things" in that explanation are called polynomials, and they have more than one piece, or term. This problem wants us to take away *all three* terms of our second polynomial.

$(9x^2 + 5x) - (4x^2 - 7x - 5)$

$5x^2 + \square x + 5$

Term ① Term ② Term ③

$(9x^2 + 5x) - (4x^2 - 7x - 5)$

$(9x^2 + 5x) - (4x^2)$ ← Term①

$5x^2 + 12x$

STEP ONE: Take away the first term. If we take away $4x^2$ from $9x^2 + 5x$, it will leave us with $5x^2 + 5x$. The $9x^2$ and the $4x^2$ are like terms.

STEP TWO: Take away the second term. This time we are subtracting a $-7x$. If we subtract a negative, it helps to think of those negative signs canceling out. We actually end up *adding* 7x on in this problem. ($5x^2 + 5x$) becomes ($5x^2 + 12x$) when we add 7 more x's.

Term ② Term ③

$(5x^2 + 5x) - (-7x - 5)$

$(5x^2 + 5x) - (-7x)$ ← Term②

$5x^2 + 5x + 7x$

$5x^2 + 12x$

Term ③

$(5x^2 + 12x) - (-5)$

$5x^2 + 12x + 5$

No like terms

$5x^2 + 12x + 5$

STEP THREE: Take away the third term. Again, we are subtracting a number that is *already* negative. Subtracting a negative means we end up adding it, so when we subtract -5, we end up with $4x^2 + 12x + 5$.

A faster way to think through this problem is to distribute the negative sign to all the things that are being taken away ($4x^2$ and $-7x$ and -5) and then combine like terms. Either way, to solve this, you end up with $4x^2 + 12x + 5$, and the book wants us to continue to the page that matches our x coefficient. For this problem, that means continuing to page 12!

ANSWER KEY FOR THIS TYPE OF PROBLEM:

Page 2: Continue to page 12
Page 5: Continue to page 10
Page 22: Continue to page 12
Page 30: Continue to page 12
Page 36: Continue to page 10
Page 40: Continue to page 12
Page 45: Continue to page 12

Polynomial multiplication (pages 9 and 11)

This is a multiplication problem, even though it looks way more complicated. When we have parentheses touching parentheses, there is a hidden multiplication sign in between. We are multiplying one thing (3x − 2) by another thing (4x + 5). You can think of these problems lots of ways (area model, FOIL, etc.). My favorite strategy is to use double distribution.

$$(3x-2)(4x+5)$$
$$12x^2 + \square x - 10$$

$$5(2x+7)$$
$$5(2x+7)$$
$$10x+35$$

When we see a problem like the one to the left, we know that we can use the distributive property to help us multiply the 5 by *both terms* inside of the parentheses.

Our problem works the same way, but we just have to distribute an extra time: once with 3x and again with the −2.

STEP ONE: Distribute the 3x. We want to multiply the 3x by both terms in the other polynomial. This will give us $12x^2$ and 15x.

$$(3x-2)(4x+5)$$
$$(3x-2)(4x+5)$$
$$12x^2+15x$$

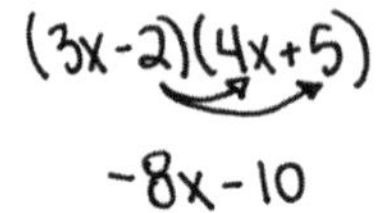

STEP TWO: Distribute the −2. Be careful here. The negative sign goes with the 2, so we need to distribute a −2! When we multiply, we get −8x and −10.

STEP THREE: Combine like terms. After distributing both the 3x and the −2, we are left with the terms $12x^2$, 15x, −8x, and −10. The 15x and the −8x terms are both regular old "x's" so we can combine them. 15 x's minus 8 x's leaves us with 7 x's, so our final answer can be written as $12x^2 + 7x - 10$. The book wants the coefficient with the x so we would continue to page 7!

$$12x^2+15x-8x-10$$
$$12x^2+7x-10$$

ANSWER KEY FOR THIS TYPE OF PROBLEM:

Page 9 (left): Continue to page 7
Page 9 (right): Continue to page 17
Page 11 (left): Continue to page 15
Page 11 (right): Continue to page 6

Visual patterns (page 26)

This problem gives us a pattern and asks for the 6th figure in the pattern. There are a bunch of strategies you can use to solve this question, and I hope you tried it in your own way!

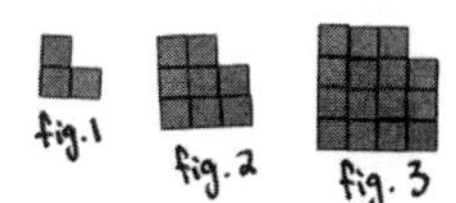

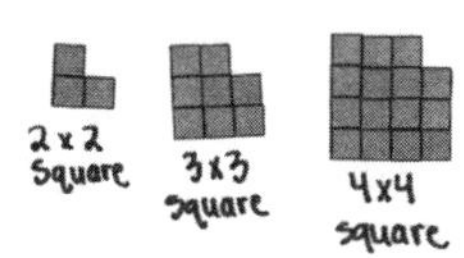

STEP ONE: Identify the pattern. The three shapes that we were given have 3, 8, and 15 blocks. This pattern isn't adding the same thing each time, but it does look like a square getting bigger. This pattern deals with an x-squared (x^2) pattern.

STEP TWO: Model the pattern with an equation. Square patterns can be modeled with $y = x^2$ equations. In this one, each side of the square looks like it is one longer than the figure number, or $x + 1$. After we create the right size square, all the figures take away one tile from the top corner. So if we take the side length $(x + 1)$, then square it, then take away the tile in the corner, we end up with $y = (x + 1)^2 - 1$.

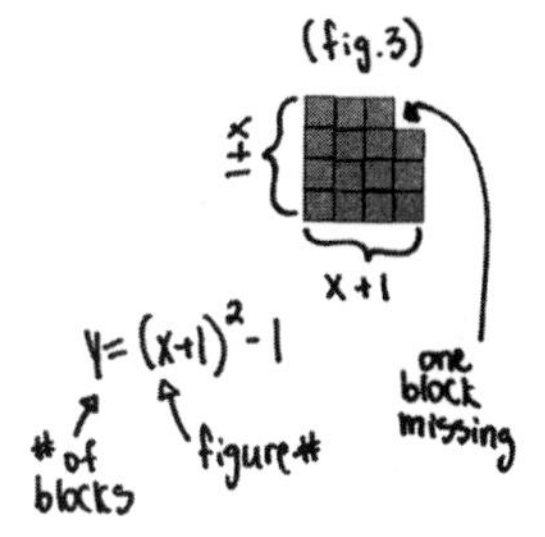

$$
\begin{aligned}
y &= (x+1)^2 - 1 \quad (x = 6)\\
y &= (6+1)^2 - 1\\
y &= (7)^2 - 1\\
y &= 49 - 1\\
y &= 48
\end{aligned}
$$

STEP THREE: Plug in 6 for x and solve. If we plug 6 (the figure that we're looking for) into the equation, it will tell us how many blocks we need. Follow order of operations and we get 48 blocks! Continue to that page.

ANSWER KEY FOR THIS TYPE OF PROBLEM:

Page 26: Continue to page 48

Finding zeroes with factoring (page 16)

We can think of this problem a few different ways. The roots/zeroes/solutions of a parabola are the places that the parabola crosses the x-axis. One way to find our roots is to graph it and look at the x-axis. We can also find those points with our algebra skills. We are looking for where our parabola equals zero, so we can rewrite our equation as $0 = x^2 - 8x + 12$. Now our equation has a bunch of numbers, but only one variable (x) so we can solve it! We just need to factor it first. Even though we only have one variable (x), it shows up in two places. x^2 and $-8x$ aren't like terms, and we can't solve for one without making the other one more complicated, so we need to change our equation from one that's in standard form ($ax^2 + bx + c$) to one in factored form [$(x+a)(x+b)$].

$y = x^2 - 8x + 12$
x = ☐
x = ☐ → ☐☐

This process is weird, but it helps to think of multiplying polynomials to understand what we are doing. Check back to page 54 of this chapter's Adventurer's Advice for a more complete breakdown, but if we multiply two binomials, let's say $(x+4)(x+5)$, we get $x^2 + 9x + 20$.

In our problem, we want to do this work, only in reverse. If we look closely at our standard form equation above, we will notice that 9 is the *sum* that we get when we add the two numbers in our factored form together: $4 + 5 = 9$. 20 is the *product* that we get when we multiply those same two numbers together: $4 * 5 = 20$. We can use that pattern on our problem too!

$(x+4)(x+5)$ Factored form
$x^2 + 5x + 4x + 20$
$x^2 + 9x + 20$ standard form

$y = x^2 - 8x + 12$
factors of 12:
12 × 1
6 × 2
3 × 4

STEP ONE: Think of all the factors of our constant. We are going to use the same pattern that we used in the $x^2 + 9x + 20$ problem. In that problem, $4 * 5 = 20$, so in our problem what numbers *multiply* to give us *twelve*? $12 * 1$ works. So too do $6 * 2$ and $3 * 4$, but which of those pairs do we pick?

STEP TWO: Pick the factors that could add up to our x coefficient. In the example problem, $4 + 5 = 9$. In our problem, we need to get to a -8. Can we get to 8 using a 12 and a 1? We could get to 13 ($12 + 1 = 13$) or 11 ($12 + (-1) = 11$), but 8 doesn't work with this pair. Looking at the other pairs of numbers, I can't get to 8 with 3 and 4 either, but 6 and 2 look more promising: ($6 + 2 = 8$). Let's pick those!

~~12 × 1~~
6 × 2 ✓
~~3 × 4~~

STEP THREE: Balance the negatives. 6 times 2 is 12, but 6 + 2 is a positive 8, not a negative one. We need to fiddle with our negatives a little bit. Let's see what happens when I make the 6 negative, but I keep the 2 positive. Negative 6 added to positive 2 is negative 4. We're closer to the −8 that we're looking for, but we actually screwed up our +12. Negative 6 times positive 2 is a negative 12, not a positive one, so we need to keep trying.

If we make both the 6 and the 2 negative, they add up to −8, *and* they multiply to a positive 12! That's the pair of numbers that we need.

$(-6) \times (-2) = +12$
$(-6) + (-2) = -8$
They must both be negative!

$0 = (x \_\_)(x \_\_)$
$-2 \quad -6$
$0 = (x-2)(x-6)$

STEP FOUR: Rewrite as a factored form equation. In this case, we found the values in our factored form equation to be −2 and −6. We just need to plug those values into the general equation of 0 = (x + ____)(x + ____), and we get 0 = (x − 2)(x − 6).

STEP FIVE: Find the solutions. This last step only works because our equation says "= 0." We are multiplying two things together and the answer is zero. If that answer has to be zero, then one of the "things" has to be zero too. Zero times anything is zero, so if we can find the place where the first thing (x − 2) is zero, then the whole equation will equal zero. (x − 2) will equal zero when x is 2, so 2 is one of our solutions. If we use the same logic on the other "thing" (x − 6), we find that the other answer is 6.

Now that we have our two solutions, we can find our page number. Put the roots together, smaller number first, to form a single number. Continue to page 26.

$0 = (x-2)(x-6)$

If this x equals 2 the whole equation equals 0

If this X equals 6 the whole equation equals 0

$X = 2$ $X = 6$

ANSWER KEY FOR THIS TYPE OF PROBLEM:

Page 16: Continue to page 26

Factoring (solve first) (pages 13, 14, 21, 23, 31, 33, and 41)

We can think of this problem a few different ways. It can be the line 11x crossing the parabola $x^2 + 28$, or we can think of it as an equation for just a parabola that we need to rearrange first. The roots/zeroes/solutions of a parabola are the places that the parabola crosses the x-axis, and we can find those points with our algebra skills.

$x^2 + 28 = 11x$
x = ☐
x = ☐ → ☐☐

$$\begin{aligned} x^2 + 28 &= 11x \\ -11x \quad & \quad -11x \\ \hline x^2 - 11x + 28 &= 0 \end{aligned}$$

STEP ONE: Rearrange the equation so that all the terms are on one side. We are looking for where our parabola equals zero, so we can rewrite our equation as $0 = x^2 - 11x + 28$.

STEP TWO: This type of equation is one that we can solve with factoring, and there is a more thorough explanation on page 56 of this chapter's Adventurer's Advice. When we look for numbers that multiply to 28, but add to −11, we end up with −7 and −4. This means we can rewrite our equation as $0 = (x - 7)(x - 4)$. We are looking for the places where the parabola is actually *equal to zero*, and those two places (our solutions) would be at 7 and 4.

$x^2 - 11x + 28 = 0$
$(x - 7)(x - 4) = 0$

If this x equals 7 the whole equation equals 0

If this x equals 4 the whole equation equals 0

$x = 7$ $\quad$ $x = 4$

Now that we have our two solutions, we can find our page number. Put the two solutions together, smaller number first, and continue to page 47.

ANSWER KEY FOR THIS TYPE OF PROBLEM:

Page 13: Continue to page 47
Page 14: Continue to page 46
Page 21 (left): Continue to page 29
Page 21 (right): Continue to page 37
Page 23 (left): Continue to page 18
Page 23 (right): Continue to page 49
Page 31: Continue to page 27
Page 33 (left): Continue to page 39
Page 33 (right): Continue to page 28
Page 41: Continue to page 38

Solving with square roots (page 8)

There are a few ways to solve this one. We could graph it or solve it with factoring. Because it has the variable (x) in only one place, we can also use our equation-solving skills here and solve with square roots.

$$x^2 + 20 = 101$$

$$\begin{array}{r} x^2 + 20 = 101 \\ -20 \quad -20 \\ \hline x^2 = 81 \end{array}$$

STEP ONE: Get rid of 20 with inverse operations. The x's are in only one place in our equation, so we can get to our answers by isolating the variable. We can get rid of +20 by subtracting 20 on both sides.

STEP TWO: Get rid of the exponent with inverse operations. The inverse operation to squaring a number is to take the square root. Square rooting an x^2 will leave us with just a normal x. When we take the square root of 81, we are answering the question: "What number, if I multiply it by itself, will give me 81?" 9 times 9 is 81, so 9 is one solution. Negative 9 times negative 9 will also give us 81, so 9 *and* −9 are solutions. The instructions on this problem tell me to disregard the −9, so continue the story on page 9!

$$x^2 = 81$$
$$\sqrt{x^2} = \sqrt{81}$$
$$x = 9$$
$$x = -9$$

ANSWER KEY FOR THIS TYPE OF PROBLEM:

Page 8 (left): Continue to page 9
Page 8 (right): Continue to page 11

Quadratic formula (page 48)

Sometimes we have quadratic (x^2) equations that don't easily factor (or equations that are impossible to factor). The one on this page has a 2 as its lead coefficient, and we can't divide the whole thing by 2 because 11/2 will give us a decimal. You can still solve this a few ways, but they will all take more time than factoring. We could graph it and see where the roots are that way. You could also solve this with strategies called "grouping" or "completing the square," but we are going to walk through this problem with the quadratic formula. The quadratic formula is helpful because it works for *any* quadratic equation, and the steps remain the same. It looks like a mess, but we just need to plug in our values and follow our order of operations.

$$y = 2x^2 - 11x - 40$$

$$x = \frac{-b \pm \sqrt{b^2 - 4ac}}{2a}$$

$$x = \frac{-b \pm \sqrt{b^2 - 4ac}}{2a}$$

$a = 2 \quad b = -11 \quad c = -40$

$$x = \frac{-(-11) \pm \sqrt{(-11)^2 - 4(2)(-40)}}{2(2)}$$

STEP ONE: Plug values into the equation. The quadratic formula uses a, b, and c, and those letters correspond to our coefficients in our equation. A is 2, B is −11, and C is −40 (make sure you catch the negatives!).

STEP TWO: Simplify the discriminant. The rest of this problem is going to involve simplifying this big mess of numbers and operations until we get our solutions. The best place to start is with the discriminant, which is the name that we sometimes give to the part of the formula under the square root symbol. −11 squared is 121, and when we multiply −4, 2, and −40, we get 320. The negatives on the 4 and the 40 cancel, and we end up adding 121 to 320 to get 441. The square root of 441 is 21.

$$\sqrt{(-11)^2 - 4(2)(-40)} \to \sqrt{121 - 4(2)(-40)} \to \sqrt{121 + 320} \to \sqrt{441} \to 21$$

$$\frac{-(-11) \pm 21}{2(2)} \to \frac{11 \pm 21}{4}$$

(+): $\frac{11+21}{4} \to \frac{32}{4} \to x = 8$

(−): $\frac{11-21}{4} \to \frac{-10}{4} \to x = -2.5$

STEP THREE: Simplify the rest of the formula. The discriminant turned into 21, and when we look at what we have left, we can continue to simplify. −(−11) is just a positive 11, and 2(2) is multiplying, so the denominator is 4. The "±" sign is weird, but all it is telling us to do is find one answer by adding (+) and one by subtracting (−). I like to split the formula into one branch with each operation, and then solve.

The book tells us to disregard the negative solution (−2.5), so we would continue to page 8!

ANSWER KEY FOR THIS TYPE OF PROBLEM:

Page 48: Continue to page 8

Discriminant (pages 27, 28, 29, 37, 38, 46, 47, and 49)

Remember what the solutions (or roots, or zeroes) mean in context of a parabolic graph. They are the places that the parabola equals zero, and where the parabola hits the x-axis.

$$y = 11x^2 - 10x + 7$$

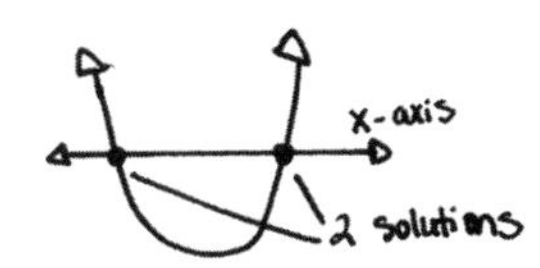

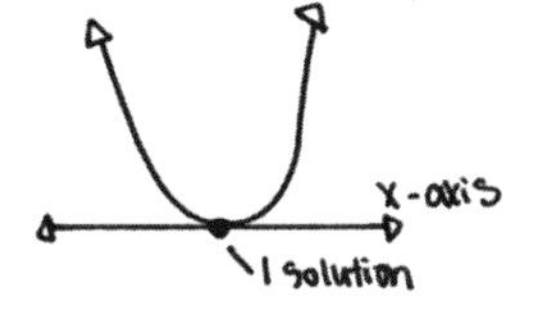

What happens when our parabola is shifted up? This parabola (at left) has only one solution.

And this parabola (at right) doesn't have any real solutions at all.

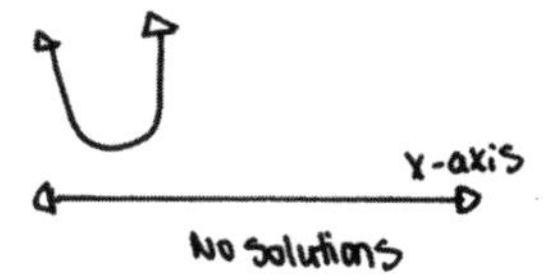

Telling these types of graphs apart (2 solutions vs. 1 solution vs. 0 solutions) can be important, and there is a sneaky way to check how many solutions a parabola has. It involves the quadratic formula, and specifically the part of the quadratic formula under the square root. This little piece of the quadratic formula is called the "discriminant," and let's see what happens when we plug the values from this problem into the discriminant.

Quadratic formula — Discriminant

$$x = \frac{-b \pm \sqrt{b^2 - 4ac}}{2a}$$

$$b^2 - 4ac$$

$$a = 11 \quad b = -10 \quad c = 7$$

$$(-10)^2 - 4(11)(7)$$

STEP ONE: Substitute values into discriminant. Just like in the quadratic formula, the values a, b, and c correspond to the coefficients in our quadratic equation. In this case, 11 (a), –10 (b), and 7 (c).

STEP TWO: Simplify the discriminant. If we follow order of operations, we want to square (-10) first. Then we multiply (-4) by (11) and (7). Finally, we subtract. Our equation will give us a discriminant of -208.

$$(-10)^2 - 4(11)(7)$$
$$100 - 4(11)(7)$$
$$100 - 308$$
$$-208$$

-208

A negative discriminant means no real roots!

STEP THREE: Evaluate. The negative that we got with our 208 on the last step is important here. The very next step in the quadratic formula would be to take the square root of that -208 number, but we have a problem. A negative under a square root is impossible (unless we get into imaginary numbers).

Remember, the square root of 25 is both 5 (because 5 times 5 equals 25) *and* -5 (because -5 times -5 *also* equals 25). What kind of number is left for the square root of -25? There is no real answer to that question! What this tells us about our parabola is that there aren't real places where the parabola crosses the x-axis. To put that more simply, this parabola doesn't have any real solutions!

If our discriminant is positive, our parabola will have two solutions, and if we get a discriminant that is exactly equal to zero (not positive *or* negative) we have a parabola with one solution!

The book tells us to follow "zero solutions" to page 5, so continue your adventure on that page!

ANSWER KEY FOR THIS TYPE OF PROBLEM:

Page 27 (left): Continue to page 5
Page 27 (right): Continue to page 44
Page 28: Continue to page 36
Page 29 (left): Continue to page 40
Page 29 (right): Continue to page 2
Page 37: Continue to page 45
Page 38 (left): Continue to page 44
Page 38 (right): Continue to page 5
Page 46 (left): Continue to page 22
Page 46 (right): Continue to page 24
Page 47: Continue to page 36
Page 49: Continue to page 30

Finding the vertex (pages 25, 32, 34, and 42)

$$y = -2x^2 + 4x + 2$$

$$(\square, \square) \rightarrow \square\square$$

The vertex of a parabola is the lowest point (if it opens upward) or the highest point (if it opens downward).

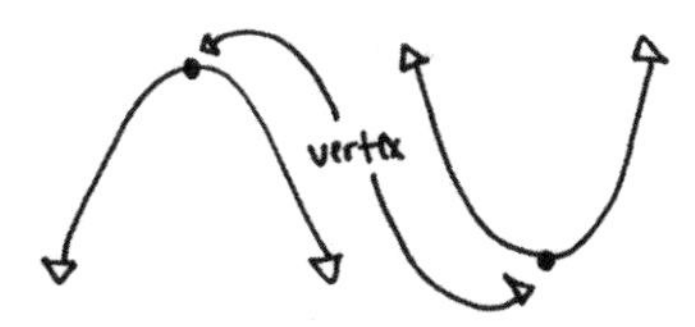

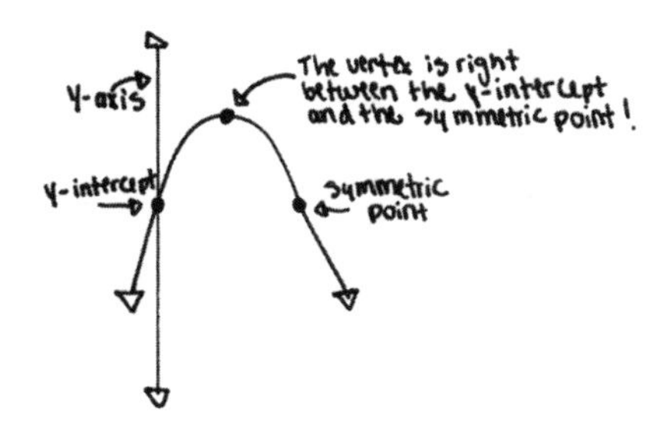

There are several ways to find this point, but we need to make sure that our answer is an (x,y) ordered pair! One of those ways works by using our y-intercept, and by remembering that parabolas are perfectly symmetrical.

STEP ONE: Find the y-intercept. For this problem, our parabola will cross the y-axis at 2.

$$y = -2x^2 + 4x + \underbrace{2}_{\text{y-intercept}}$$

STEP TWO: Plug the y-intercept in for y. After we know one place where our y-value is 2—the y-intercept, at the point (0,2)—we want to find the other place our parabola has a y-value of 2. We can set our parabola equal to 2, and in the next few steps we will solve to find exactly where our parabola reaches up to a value of 2.

y-intercept = 2

$$y = -2x^2 + 4x + 2$$
$$2 = -2x^2 + 4x + 2$$

STEP THREE: Solve for the y-intercept's symmetrical point. By subtracting 2 from both sides and factoring out an "x," we can rearrange our equation to say 0 = x (−2x + 4). The solutions to this equation will show us both places where the parabola is equal to (has a height of) 2. In this problem, those two places are at zero (that one is the y-intercept) and at the point (2,2).

$$2 = -2x^2 + 4x + 2$$
$$-2 \qquad\qquad -2$$
$$0 = -2x^2 + 4x$$
$$0 = x(-2x + 4)$$

If this x equals 0 the whole equation equals 0 — If this x equals 2 the whole equation equals 0

$x = 0$ → (x, y) → $(0, 2)$ (height of 2)

$x = 2$ → (x, y) → $(2, 2)$ (height of 2)

$x = 1$

$y = -2x^2 + 4x + 2$

$y = -2(1)^2 + 4(1) + 2$

$y = -2(1) + 4(1) + 2$

$y = -2 + 4 + 2$

$y = 4$

$x = 1 \quad y = 4$

(x, y)

$(1, 4)$

STEP FOUR: Find the point in between our y-intercept and its symmetric point. This parabola has two points, (0,2) and (2,2), that are the same height. Because we know that a parabola is perfectly symmetrical, the vertex has to be right in between these two points. Our vertex is going to be right in between 0 and 2, at 1.

STEP FIVE: Plug the vertex value into the parabola equation. We're close! We know that the vertex has an x-value of 1, but we need to find how high up or low down our vertex is. In order to do that, we will plug 1 into our parabola equation and use order of operations to solve. We find that when our x-value is 1, our y-value is 4. Our vertex is at the point (1,4).

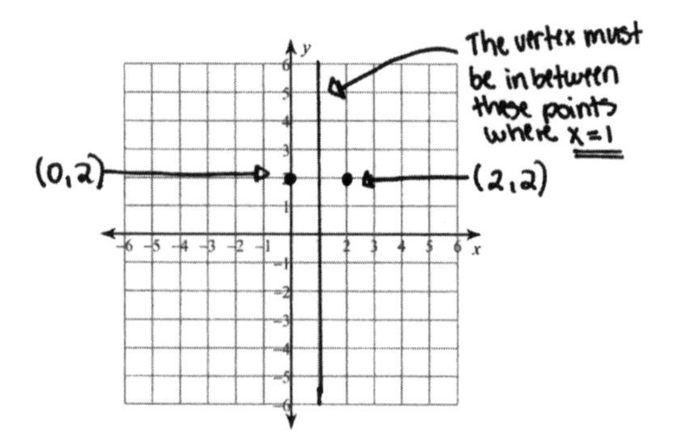

Turn (1,4) into the page number 14, and continue to that page!

AUTHOR'S NOTE: There is a great shortcut we can use on problems like this, and it comes from the quadratic formula. Turn back to page 60 of this chapter's Adventurer's Advice for a reminder of quadratic formula, but the "-b/2a" portion of the quadratic formula will give us the x-value between the y-intercept and the symmetric point. In this problem, "b" is +4 and "a" is −2 (just like we would find if we were using the quadratic formula). When we substitute those values into -b/2a, we get −4/2(−2), which reduces to 1. We still need to plug 1 into the equation (Step Five), but this shortcut will save us quite a bit of time!

ANSWER KEY FOR THIS TYPE OF PROBLEM:

Page 25 (left): Continue to page 14
Page 25 (right): Continue to page 21
Page 32 (left): Continue to page 13
Page 32 (right): Continue to page 33
Page 34 (left): Continue to page 43
Page 34 (right): Continue to page 23
Page 42 (left): Continue to page 31
Page 42 (right): Continue to page 41

Quadratic system (substitution) (pages 6, 7, 15, and 17)

$y = x^2 - 4x + 5$
$y = -2x + 8$

The parabola $y = x^2 - 4x + 5$ crosses the line $y = -2x + 8$ in two places. One of those places is (−1,10) and the other is . . .

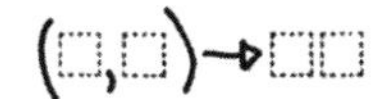

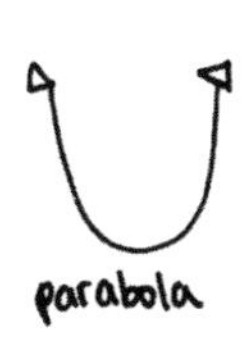

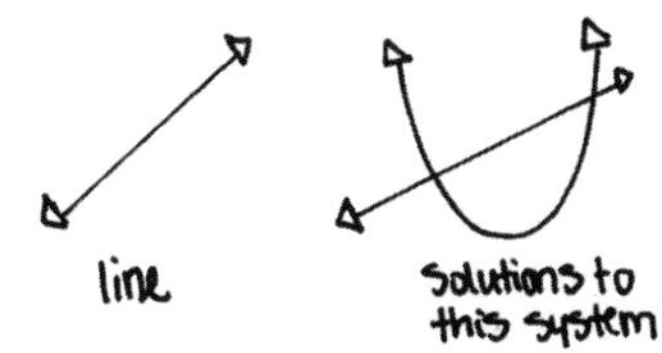

This problem is looking for a solution to this system. The solution is the place where these lines cross and these equations are equal. Because we are dealing with a parabola and a line, this problem is going to have up to two places where the line and the parabola meet.

Before we start, it helps to remember that the places these lines are going to cross is going to be an (x,y) ordered pair, so we need to find a solution in that format. For this problem, both equations are solved for y, which makes it easy to solve with a strategy called *substitution*.

$y = x^2 - 4x + 5$
$y = -2x + 8$
$-2x + 8 = x^2 - 4x + 5$

STEP ONE: Substitute. The second equation tells us that y is equal to $-2x + 8$. Those two quantities are exactly the same, so we can look to the first equation. Instead of y in the first equation, we can write in the thing that y is equal to: $-2x + 8$. This gives us the new equation: $-2x + 8 = x^2 - 4x + 5$. This helps us, because now we only have one variable and we can solve!

STEP TWO: Get all the terms to one side. This problem involves a quadratic equation, so we can't solve it with inverse operations. Instead, we need to factor, and before we can do that, we need to get all the terms to one side by adding 2x on both sides and subtracting 8 on both sides. The resulting equation is $0 = x^2 - 2x - 3$.

$-2x + 8 = x^2 - 4x + 5$
$+2x \qquad\qquad +2x$
$8 = x^2 - 2x + 5$
$-8 \qquad\qquad -8$
$0 = x^2 - 2x - 3$

$0 = x^2 - 2x - 3$ → factor!
$0 = (x+1)(x-3)$

If this x equals −1, the whole equation equals 0

If this x equals 3, the whole equation equals 0

STEP THREE: Solve the equation using factoring. For a more detailed breakdown, check out page 58 of this chapter's Adventurer's Advice. The numbers that *multiply* to give us −3 but *add* to give us −2 are −3 and a positive 1. That means that we can rewrite this equation in factored form as $0 = (x + 1)(x - 3)$. The places where this equation is *actually equal to zero*, like the equation says, are at positive 3 and negative 1.

STEP FOUR: Plug in our x-values and solve for y. Even though we have already done a bunch of work, we still have a few steps left. From here, we need to find where our lines are when x is equal to −1 and at +3. We do that by plugging *our x values* for x into one of the equations. It doesn't actually matter which equation we pick, because these are the *only places* where the two lines are at the same place on our graph! I'm going to use $y = -2x + 8$, but you can use the other equation if you want. Follow order of operations, and we find that when we plug in −1, we get 10, and when we plug in 3, we get 2.

$x = 3$
$y = -2x + 8$
$y = -2(3) + 8$
$y = -6 + 8$
$y = 2$

$x = -1$
$y = -2x + 8$
$y = -2(-1) + 8$
$y = 2 + 8$
$y = 10$

$x = 3$, $y = 2$
(x, y)
$(3, 2)$

$x = -1$, $y = 10$
(x, y)
$(-1, 10)$

STEP FIVE: Put your x and y values together to find the solutions. We have already done all the math, now we just need to plug in our numbers to get two (x,y) ordered pairs. The first ordered pair is (−1,10), but this problem told us to disregard that point. The other one is at (3,2), and when we put those together to create a page number, we get 32. Continue to that page!

ANSWER KEY FOR THIS TYPE OF PROBLEM:

Page 6: Continue to page 32
Page 7: Continue to page 34
Page 15: Continue to page 42
Page 17: Continue to page 25

CHAPTER 4

You take a deep breath. *Dregg Corp has created portals? And are using them to commit crimes? What a weird day . . .*

All you intended to do was return a lost wallet (and collect the reward money) but instead you've unlocked Dregg Corp's deepest, darkest secrets.

You consider your mom, and your friends Hazar and Taylor, who you've been messaging and asking to help. Have you endangered them? How much will your new knowledge be worth to Dregg Corp? And what will they do to keep their secrets hidden from the public?

You are deep in thought when soft whirring sounds bring you back to reality. The control panels for the first and the second portal softly purr as they each eject a small, metallic card.

You grab one of the cards and turn it over in your hand. It's a heavy metallic one with a Dregg logo embossed on it, *JUST like the one you found in Doctor LaBella's wallet*!

You dig the wallet out of your pocket to compare the card from the control panel with the one you found on the street. They are the same size and the same weight, but the Dregg logos are different colors. The card from Doctor LaBella's wallet has a blue "D" and the 'D' on the card from the control panel is purple. There is a third portal, and you now have three key cards.

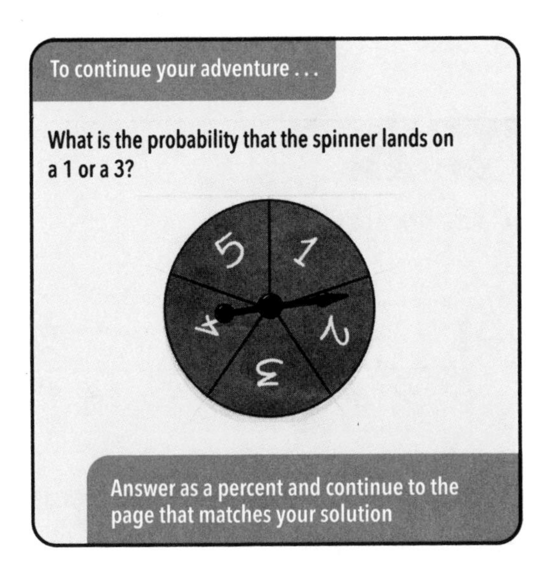

See page 51 of Adventurer's Advice Chapter 4 for help.

You should have arrived here from page 42 OR page 45

Big red button? No thank you. You and the doctor push deeper into the maze.

As you round a corner, you see two giant furry shapes up ahead. They have snow-white fur, long flesh-colored tails, and beady red eyes. These are the biggest rats you have ever seen. They scurry over to investigate the interlopers in their maze.

"Hey there, Ratigan," says the doctor, scratching one creature behind its ears. "Can you show us the way out of here?" The giant rat smells her hand and nuzzles up against her leg.

The creature turns and leads you through a series of twists and turns in the maze, and before long you emerge into a cavernous chamber where it looks like the rats have been living. They have built nests against the back wall with insulation and scraps of wood and cardboard. You also see a tunnel that the rats have gnawed into the wall with their big, blunt front teeth. Ratigan leads you into the tunnel in the wall.

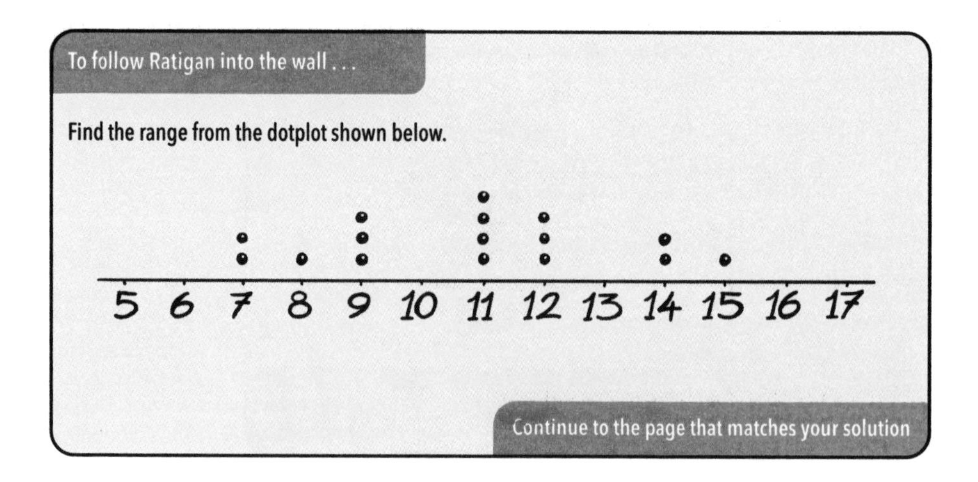

See page 61 of Adventurer's Advice Chapter 4 for help.

You should have arrived here from page 42 OR page 45

The red button is about five inches across. When you press it down, a hatch in the wall opens and a giant round hunk of Swiss cheese tumbles out onto the floor.

You hear some skittering noises from down the corridor, and two giant rats emerge from the shadows, each one standing four feet off the ground.

They are bright white in color with beady red eyes. They jostle for position as their big blunt teeth tear into the hunk of cheese.

Doctor LaBella approaches the creatures and runs her hand through the fur behind the ears of the first rat. "Hey, Ratigan!"

She grabs the scruff of the neck of the giant creature and swings her left leg over its back. "What do you think? Can you get us out of here?"

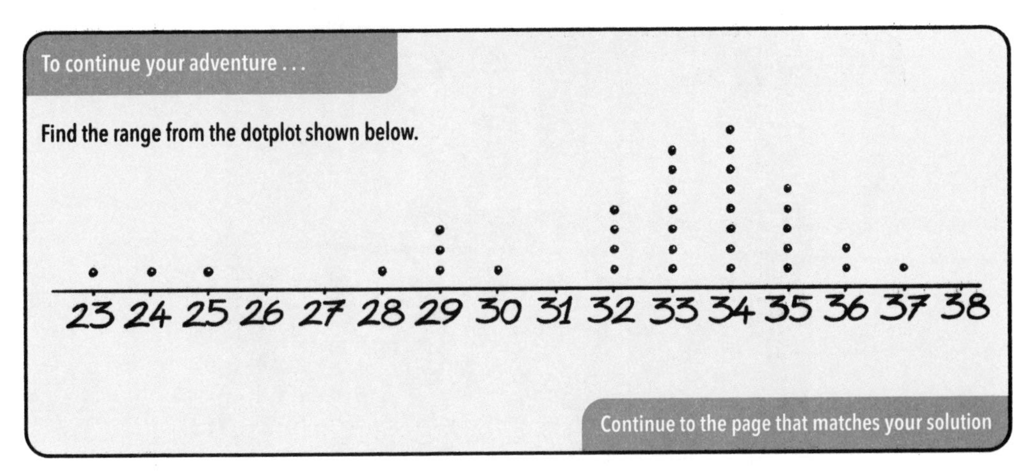

See page 61 of Adventurer's Advice Chapter 4 for help.

You should have arrived here from page 36

The museum is three stories tall, and the window-washing platform wasn't built for speed. Now that it has power, it begins to descend . . . *slowly.*

You are still twenty feet off the ground when the low rumble of the generator quiets and the platform stops with a jolt. *The guards must have turned off the generator.* You are stuck, halfway between the roof and the ground.

The front of the museum has large windows set between blocky limestone pillars. You climb off the platform and onto a windowsill, hugging your body close to the limestone façade.

You nervously try to scoot along the windowsill, spying an iron drainpipe at the end of the building.

Unfortunately, your night of chaos has robbed you of your usual balance and grace—twenty feet isn't too far, but you may not survive this fall.

D

O

W

N

you go!!!

The End

You should have arrived here from page 50

Two Dregg security guards pull open the door to the fifth floor, and they are immediately met with a stampede of manicured animals.

The two guards at the door are knocked back into the stairway and into the other guards looking to apprehend you.

You manage to infiltrate the herd and stay with the animals as they exit the fifth-floor lab and head down the final flights of stairs to the first floor.

The chaos from the fabulous creatures is providing a perfect cover for your escape, and in no time you have made it to the front door of Dregg Tower. The animals are free, and so are you.

The End

You should have arrived here from page 44

What now? you wonder to yourself. You look around and notice a fire hose coiled neatly in a compartment on the wall. You open the red door and pull out the heavy metal nozzle. *Is this hose long enough?*

You pull out an extra length of hose and cross the hall to a window that faces the street. You wind up and launch the heavy nozzle at the window. The window shatters, showering the ground below in a cascade of broken glass. The spool the hose is wound around spins wildly as the nozzle falls toward the ground. After a few moments, it jolts to an abrupt stop, having run out of hose.

You look out the broken window, and to your relief, the nozzle of the hose is dangling a few short feet above the ground. You grab the hose with both hands and swing your feet out through the broken window.

After a few minutes of perilous downclimbing, you reach the ground and escape into the cool night.

Should you have done more to rescue Doctor LaBella from her hypnosis? Or at least returned her wallet? And what about all the evil deeds you saw at Dregg Corp?

Best not to stick my nose into other people's business, you decide as you make your way home.

You don't even notice the raccoon wearing a backpack, trailing you from a distance, following your every move.

The End

You should have arrived here from page 16

The doctor is too deep in her hypnotic haze for you—or anybody else—to save her. The only thing for you to do now is to get out of here! Your feet are flying down the hall, and you leap through the portal back onto the eleventh floor of the Dregg Tower.

You immediately hit the OFF button on the control panel for the portal, and it sputters out before the doctor can catch up with you. You aren't in the clear yet. An alarm is ringing somewhere in Dregg Tower, and you'll have to move quickly. *Security is no doubt closing in fast.*

You dash out into the foyer just outside the lab.

There is an elevator, but you can see from the display that it is at the eighth floor and climbing. If the elevator has a Dregg security detail in it, you don't want to be here when they arrive. *Better to take the stairs down,* you think to yourself.

To take the stairs down . . .

Use the two-way frequency table below showing some of the stolen jewels and gems from the museum. What percent of the stolen jewels are sparkly?

	Sparkly	Shiny	Total
Jewels	9	51	60
Gems	14	16	30
Total	23	67	90

Continue to the page that matches your solution

See page 62 of Adventurer's Advice Chapter 4 for help.

You should have arrived here from page 2

The rat tunnel is only four feet tall and you have to crouch low to the ground to move through it. It smells terrible. You pass through a series of gnaw-holed walls, and when you pop out, you are in a small atrium with a vending machine. The rats have broken the glass and have been eating the chips and cookies inside. You pull out a bag of cheese crackers and rip it open for your rodent tunnel guides.

A sliding door opens to a small deck outside the second floor. You open the door and walk outside with the doctor. A drainpipe is running from the deck down to the ground beside the tower, and the two of you use it to slide down. It's wet, but you make it to the ground safely.

There are no clouds, and the moon provides some light as you and Doctor LaBella walk across Dregg Plaza and into the night.

"This is where we part ways," the doctor says. "And unfortunately you can't tell anyone what's happened here. But don't worry, we'll meet again. In the meantime, can you call your mom? She's absolutely panicked about where you are."

Right, you think, pulling your cell phone out of your pocket.

"What do I tell her?!" you demand. But the doctor has vanished and you're all alone.

You should have arrived here from page 27

You return to Dregg Tower with Levi Dregg.

He gives you a desk in the lab near the portals, and you get a jumpsuit, a radio, and your own security card for your new job as a criminal.

After a few weeks working with Levi after school, you are really getting the hang of things, and you're even in line for a promotion. *Assistant Thievery Manager!* You help rob jewelry stores, banks, and pharmacies, and you sell the loot online. You even rob a few birthday parties. Video games resell well, and the delicious cake is an extra perk.

You wonder what happened to Doctor LaBella. Her notes and discoveries are all over the lab. One day you find a moment to ask another researcher about her, but he just gives a low whistle and tells you he thinks she had a breakdown.

"She was a rising star, brilliant mind," he says. "But one day she just vanished."

That night on your way home from work, a man with short blond hair stops you on the street. "You don't know me, but I know you. My name is Officer Lawrence, and I have a few questions for you."

The End

You should have arrived here from page 28

You approach Doctor LaBella. She looks peaceful in her hypnotized state, but you have some questions.

"Doctor LaBella, is that you?"

"Dregg is family. LaBella is Dregg."

Not off to a great start.

Before you can ask her another question, she blinks. It's like she recognizes your voice. "You are the one with my wallet," she says.

She's right, and it startles you. "How . . . how do you know that?"

"That answer is complicated," Doctor LaBella replies. "How much particle physics do you know? These portals—they send you through space *and* time." She pauses for a moment, and you can see her trance beginning to break and life returning to her dazed facial features. "You don't remember now, but we've met before."

The doctor is slowly returning from her hypnosis and continues her explanation. "Dregg was going to leave me here to take the blame for the museum robbery, and they know you have been poking around. They will be sending their people to find you soon, and we have to get out of here before they show up. We're going to need to stick together if we are to make it out alive."

CHAPTER 4 GOAL: **Escape with Doctor LaBella!**

You help Doctor LaBella up from her chair. You don't yet know if you can trust her, but you don't really have a choice. Her left hand is bleeding a bit, and it must be the source of the bloody trail in the hallway. You help her clean her wound quickly. Time to move!

To continue your adventure . . .

Find the median of the data set shown below.

62, 37, 56, 40, 15, 22, 65, 42, 25, 47

Continue to the page that matches your solution

See page 54 of Adventurer's Advice Chapter 4 for help.

You decide getting away from the guards quickly is the most important thing. You charge back up to the sixth floor and throw open the door.

The sixth floor is dark, and your feet sink a few inches into the floor. *It smells like chicken soup in here*, you think. You reach behind the door and turn on the lights.

The sixth floor is one giant room, and it looks like Dregg is using the space as an enormous petri dish. The petri agar is the texture of jelly, and it is coming up to your ankles.

You slosh through the room and come across dozens of pulsating blobs in one corner, each one probably two feet long and a foot in diameter. Upon closer inspection, each blob appears to be a single, enormous cell. You can see nuclei, the mitochondria, and other giant organelles. One of the cells is in the process of dividing, and you watch as it separates into two, slightly smaller, daughter cells.

The doors behind you burst open, and the Dregg security guards pour into this supersized science experiment.

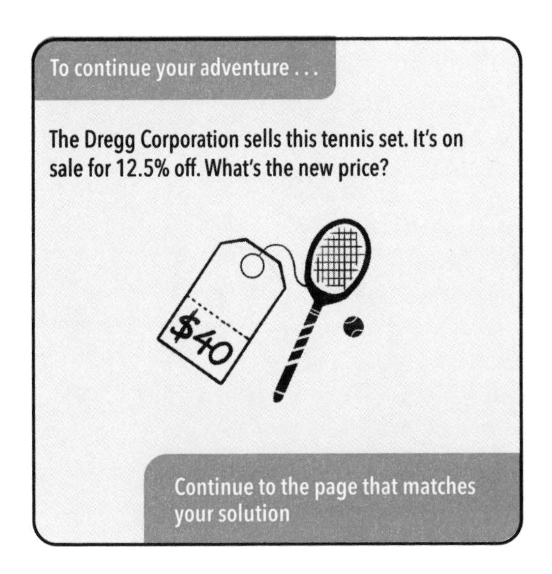

See page 52 of Adventurer's Advice Chapter 4 for help.

You should have arrived here from page 48

You take off your belt, sling it over the steel cable, and wrap it around your wrists. You take a deep breath, push off the building, and head—

STRAIGHT INTO BRIGHT LIGHTS!

Your hands are busy with the belt, so all you can do is squint as you finish your descent to the park below. You drop the last couple of feet to the grass, and when you get to your feet, you are surrounded by TV cameras and reporters shouting out your name. "How does it feel to be the one who brought down Dregg?" shouts one reporter, thrusting a microphone into your face.

Another reporter interrupts, "What do you say to all the people out there calling you a hero?"

You weren't prepared for all this attention, and you blush at the flattering questions. Through the crowd, you can see a few familiar faces. Hazar, Taylor, and your mom are behind the reporters, waving at you.

You give the cameras a big cheesy smile and break away to join your mom and friends.

"How did you all *find* me?" you ask. "I have so much to tell you!"

"Well," responds Taylor, "you must have sat on your phone weird. You've been live-streaming audio from your pocket for the last few hours. It's how these reporters found you too." Your face turns red in embarrassment. *Did you say anything stupid?*

Taylor continues. "It's actually a good thing though. You exposed all the shady stuff going on at Dregg. They're in *real* trouble."

"Let's get out of here," says Hazar with a smile. "We have math homework to do tonight."

The End

You should have arrived here from page 18

You walk into an empty cafeteria. Muted light comes in through bubble windows, the kind you'd see on the side of a submarine. You've eaten here before when you visited the museum. All the menu items have bad artist puns in the names. Rembr'ants on a log. Van Gogh't cheese salad. Andy Warh'almonds.

Uggh. *So corny.*

You walk the length of the room, wondering what to do. Doctor LaBella needs your help, but this museum is a big place. She could be anywhere. And even though the museum seems empty, you definitely aren't in here alone.

You check your cell phone. No service. Hours have passed since you found the wallet. Your mom has probably really panicked by now. You lean your forehead against the glass case of Frida Kahlo'flower bites and sigh.

"Need some help, youngster?" asks a voice. You look over and see a janitor wheeling a mop bucket. His name tag says Mr. Lucky. He looks familiar, but you can't quite place him.

To ask Mr. Lucky's advice . . .

Find the mean (average) from the data set below.

38, 51, 19, 44, 60, 46

Continue to the page that matches your solution

To say you're fine and head to the Egyptian Room . . .

Find the mean (average) from the data set below.

35, 20, 17, 21, 43, 32

Continue to the page that matches your solution

See page 56 of Adventurer's Advice Chapter 4 for help.

You should have arrived here from page 3

You are uneasy at first, but the creatures seem to calm after their cheesy feast.

"That one's name is Felix. He's a real prankster!" says Doctor LaBella.

Prankster? What does that even mean? You swing your leg over Felix's back and grab onto the scruff behind his neck.

"Hi-yah!" shouts Doctor LaBella as she spurs on Ratigan with her heel.

The rats take off at a gallop through the maze, eventually reaching a wide staircase leading down to the first floor. It takes all of your strength to hang on as the rats dash down the stairs and around a corner on the first floor.

The rats blast past the guard and burst through the double doors into the dark night outside, leaving the tower behind.

You've escaped! What happens next? You have to go where the rats take you, for now at least.

The End

"Code Scarlet. Repeat: Code Scarlet. Obtain the intruder at all costs," a robotic voice says over the intercom. It *must* be about you, and now you know you must avoid Dregg security guards no matter what it takes.

Fast as you can go, you tell yourself as you rush down the stairwell of Dregg Tower from the lab on the eleventh floor.

Eighth floor . . . seventh floor . . . sixth floor . . .

At the landing between the sixth and fifth floor, you steal a peek through the handrails to the stairs below. There is a flurry of activity as guards dressed in black come up toward you.

You could continue to the fifth floor, but beyond that, the stairs are no longer an option.

You could also turn back and try to find an escape route on the sixth floor.

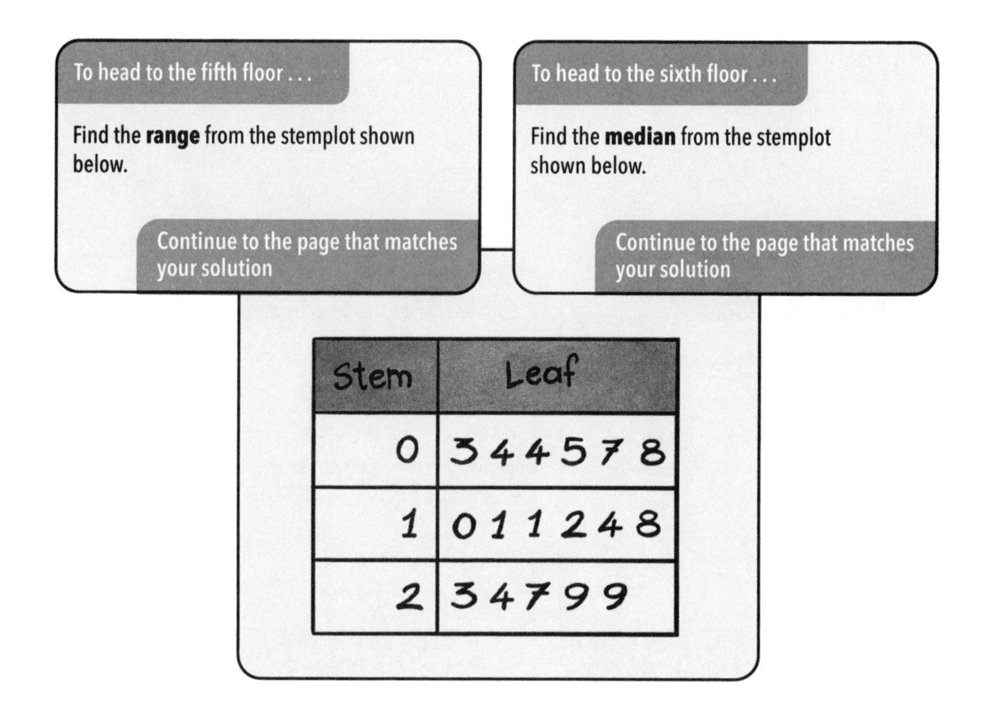

Stem	Leaf
0	3 4 4 5 7 8
1	0 1 1 2 4 8
2	3 4 7 9 9

See pages 55 and 60 of Adventurer's Advice Chapter 4 for help.

You should have arrived here from page 28

You hurry to the fountain in the Egyptian Room and cup your hands to fill them with water. You pause for a moment to take in the beauty of the fountain. A sandstone carving of Osiris sits in the middle of it, and the basin is lined with small, bright blue tiles. The surface of the water is so reflective and smooth, it seems like a shame to disrupt it.

In the corner of your eye, you spot something reflected in the perfect blue water, and you leap out of the way just in time.

The doctor's chair flies inches from your head and lands in the fountain with a splash. "Dregg is family. Dregg is power." It's the doctor. Still clearly feeling the effects of her hypnosis.

You shout to her and splash her with water, but her chanting grows louder. "*Dregg is family! Dregg is power!*" The doctor has picked up a potted plant and is walking toward you menacingly. "*DREGG IS FAMILY. DREGG IS POWER.*"

CHAPTER 4 GOAL: **Escape this museum!**

Try as you might to get through to the doctor, she is lost in this hypnotic state, and her chanting is going to bring back security soon. You have to move, and fast. Do you leave through the portal and take your chances in Dregg Tower, or do you want to escape through the museum?

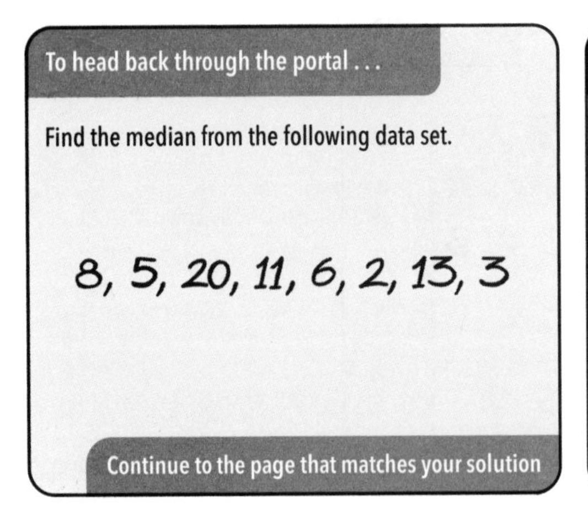

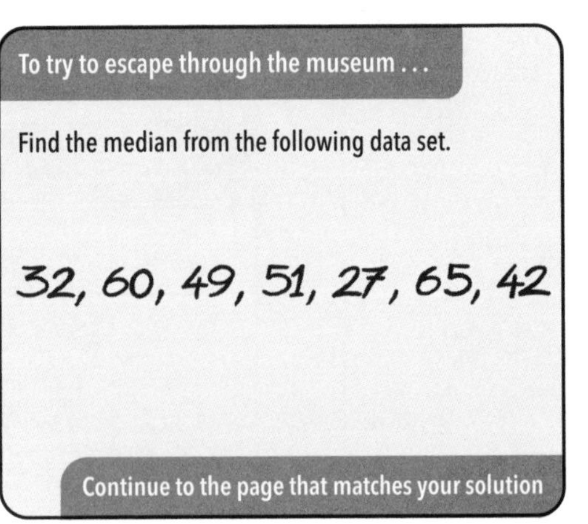

See page 54 of Adventurer's Advice Chapter 4 for help.

You should have arrived here from page 24

When you step out of the elevator into the expansive garage, you realize the cars down here are not what you expected.

They're not really *cars*.

"This level of the garage is where Dregg keeps all their experimental vehicles," Doctor LaBella explains as you look around in awe. There are motorcycles, speedboats, helicopters, even a train engine.

"Either of these two will work," says the doctor, motioning toward a tank and a hovercraft parked side by side a short walk down the garage. "Which one should we take?"

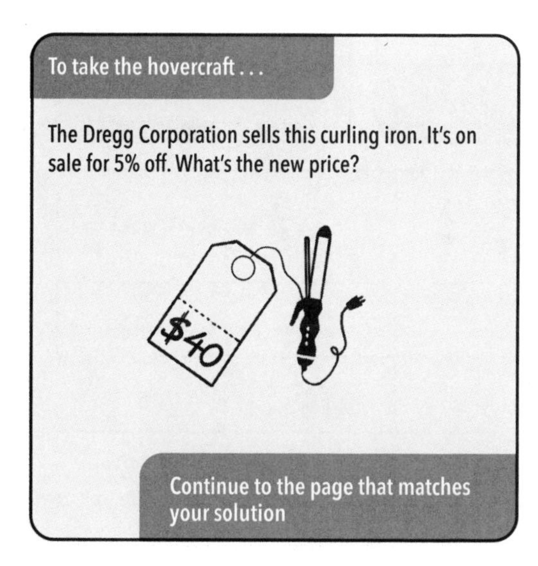

See page 52 of Adventurer's Advice Chapter 4 for help.

You should have arrived here from page 40

The portal whisks you through the ether, and you land with a thud in a dimly lit hallway. Beyond the swirling plasma of the portal, the hallway opens up into a cavernous atrium with a few large displays on the walls. It's the art museum on the other side of the city, and you can tell immediately that a lot of the artwork is missing. Looks like Dregg has already raided this place. *This explains the stacks of paintings in the lab. They must have been stolen from this museum!*

A trail of small droplets of blood leads down the hallway away from the atrium. Doctor LaBella is nowhere in sight.

You've been in the museum before on school trips, and weekends with your parents when you were younger. This hallway leads to a few of the museum's most popular exhibits. The Rare Art wing was always your mom's favorite. Dregg has no doubt raided the art from that gallery already, but maybe that's where you will find the doctor. The Egyptian Room is down this hallway too, just past the cafeteria. Where should you start your search?

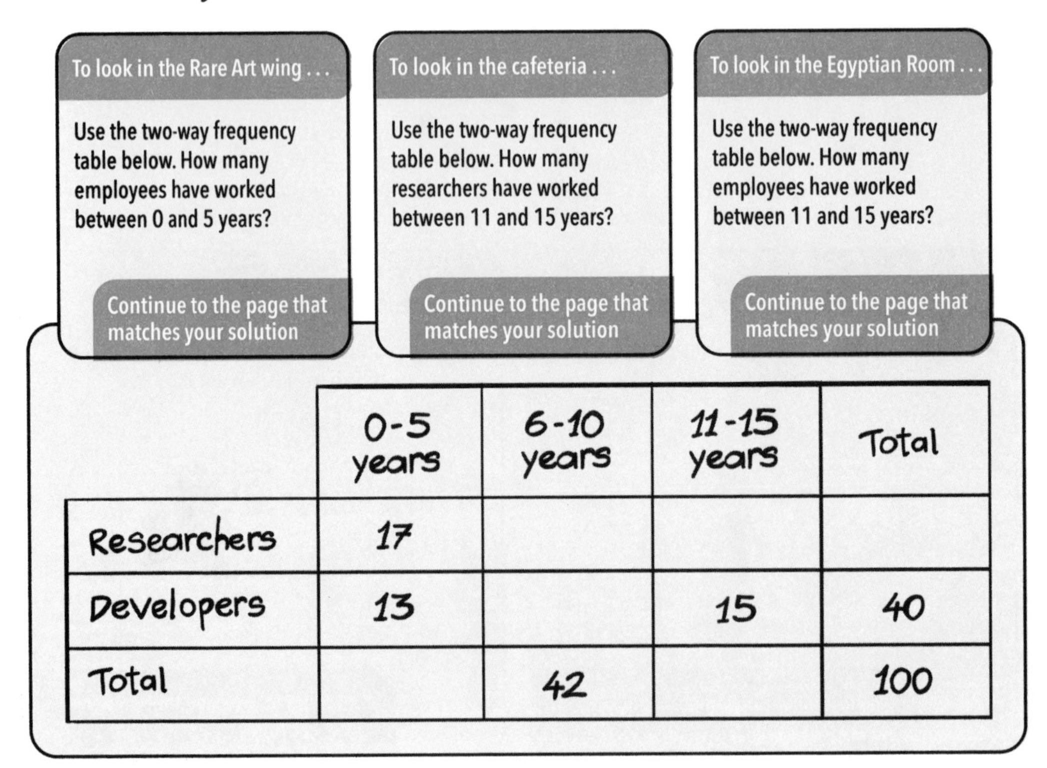

	0-5 years	6-10 years	11-15 years	Total
Researchers	17			
Developers	13		15	40
Total		42		100

See page 57 of Adventurer's Advice Chapter 4 for help.

You and Doctor LaBella turn left at the first fork and walk down the makeshift corridor. There are scraps of long gray hair on the ground, and you notice scuff marks at hip height on the wall.

"So when you said this is the big rat maze, are the rats, like, this big?" you ask the doctor, holding your hands a foot and a half apart.

"Bigger," she responds. "Way bigger. They do psychiatric and memory experiments, and they want rats with human-size brains."

You shudder as you imagine the size of rat that would have a brain that big. Hopefully your imagination is as close as you will get to seeing these rats.

The two of you reach another fork in the maze and must again turn left or right.

See page 52 of Adventurer's Advice Chapter 4 for help.

You should have arrived here from page 41

The private elevator is smaller than the service elevator that you took to get to the eleventh floor, but it moves much faster. As soon as the doors slide shut, you and Doctor LaBella are barreling down toward the second floor.

The doctor already has her ID card, but you still have her wallet. You rifle through your pockets and hand it back to her. It feels good to be rid of the thing. It has almost gotten you killed countless times today. "Hey, thanks!" she says, beaming. As she takes the wallet and slips it into her jacket pocket, the elevator doors slide open. The second floor is dimly lit and not full of offices and cubicles as you had been expecting.

You each step off the elevator and look around quizzically. The walls look clumsy, like they have been put up in a hurry. These clumsy walls form a hallway that splits before you: to the left or to the right.

"Uh-oh," says Doctor LaBella. "I forgot they moved all the corporate offices to the building across the plaza. Now this floor is . . ." Suddenly a section of wall slams down behind you, blocking you from returning to the elevator.

". . . where they put the big rat maze. Looks like we have to find our way out."

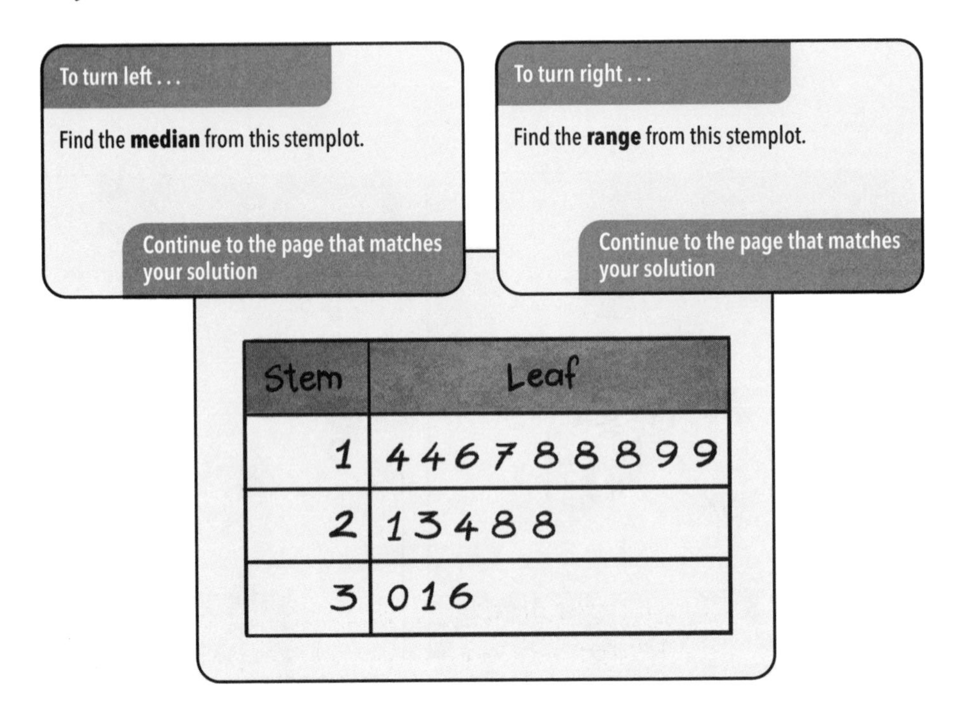

Stem	Leaf
1	4 4 6 7 8 8 8 9 9
2	1 3 4 8 8
3	0 1 6

See pages 55 and 60 of Adventurer's Advice Chapter 4 for help.

You should have arrived here from page 25

The museum roof is pebbled with small rocks stuck in tar. It makes a scuffing sound as you run toward the helicopter.

You reach it and pull open the driver-side door. *Is it called the "pilot side" door on a helicopter?* You climb inside and look around for keys, or a parachute, or anything that you could use to get to the ground below the museum.

In the rear storage area of the helicopter, you can see what you first think must be a harpoon gun. You pick it up and notice a spool of steel cable and the wide barbs at the end of the harpoon. *It's a grappling hook!*

You jump from the helicopter and run to the edge of the museum to look for a target.

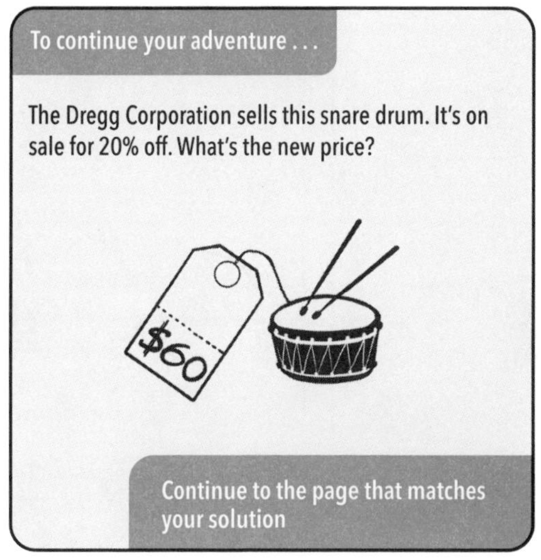

To continue your adventure . . .

The Dregg Corporation sells this snare drum. It's on sale for 20% off. What's the new price?

Continue to the page that matches your solution

See page 52 of Adventurer's Advice Chapter 4 for help.

You should have arrived here from page 20

Your only memory of being in a maze is the one corn maze you and your cousins did when you visited them out in the country, wayyyy back when you were five or six years old. You got lost for hours. NOT your ideal vacation activity and NOT something you want to repeat.

"So the big maze," you say to the doctor. "It's, uh . . . difficult?"

"Extremely," Doctor LaBella whispers to you out of the side of her mouth. "When I called this the big rat maze, I didn't mean it's a BIG MAZE for rats. I meant it's a maze for BIG RATS. Dregg scientists develop psychiatric and memory experiments, and the rats we use are *enormous*."

Well, that doesn't sound great, you think.

The two of you reach another fork in the maze and must again turn left or right.

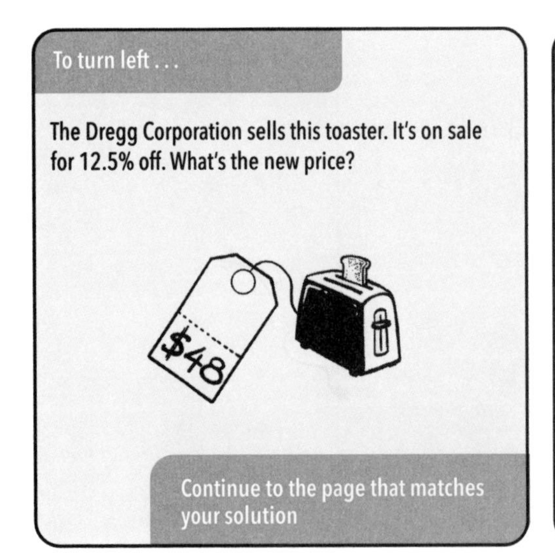

See page 52 of Adventurer's Advice Chapter 4 for help.

You should have arrived here from page 38

The hovercraft rumbles down the road, taking up two full lanes. Two Dregg security cars are in pursuit with lights flashing, and it won't be long before they catch up. The hovercraft is too slow to outrun much of anything.

Doctor LaBella takes a sharp right turn over a curb and through a grassy field. The Dregg security detail is still chasing you, and they are only a few hundred feet behind you now.

"C'mon . . . c'mon . . ." whispers Doctor LaBella to nobody in particular. As the hovercraft passes over a hill, her plan comes into focus. The wide river that runs through town is just ahead.

The hovercraft drifts between a park bench and a stand of oak trees, then careens off a five-foot ledge and into the river below. Well, *onto* the river below. One of the security cars follows you off the ledge and immediately sinks in the murky river water. The doctor steers the craft downriver and into the night.

"Lost them!" says the doctor triumphantly. "Hey, do you have a cell phone?"

"I do," you tell her, wide-eyed.

"Call up Channel 4. I'd say we have a newsworthy story to tell!"

The End

You should have arrived here from page 41

"I don't want to spend another second in Dregg Tower," you confess to the doctor. "Can we go straight to the parking garage?"

"Absolutely!" she says, in a cheerier voice than you'd expect. She's acting like you're going out for lunch together, not barely escaping with your lives.

As the elevator descends to the garage, you rifle through your pockets and hand the doctor her wallet. "Hey, thanks!" she says, eyes beaming with delight. "I'm one punch away from a free sandwich!" As she takes the wallet and shoves it into her jacket pocket, the doors slide open. The garage is lit by blue, incandescent light, and the floors and walls are dull gray cement.

To continue your adventure . . .

Find the range from the stemplot shown below.

Stem	Leaf
1	8 9
2	0 2 3 3 5 6 7 8 9 9
3	0 2 4 5

Continue to the page that matches your solution

See page 60 of Adventurer's Advice Chapter 4 for help.

You should have arrived here from page 49

The metal rungs of the fire escape end at a heavy metal covering. You undo the latch before pushing it open. You can hear the Dregg security detail below in the museum, looking for you.

The night air feels cold against your face, and as soon as you have climbed onto the roof, you latch the door shut behind you. *That should slow them down,* you think to yourself.

Across the roof you can see a helicopter on the helipad. *I can't fly a helicopter, right? Like, those are complicated.*

Maybe you're brave enough to try, after all you've been through tonight?

There is also a platform that the window cleaners use to clean the tall glass exterior of the building. It is hanging just off the side of the building to your left, where the cleaners left it at the end of their day. Might also be an escape route that way, if you want to play it safe.

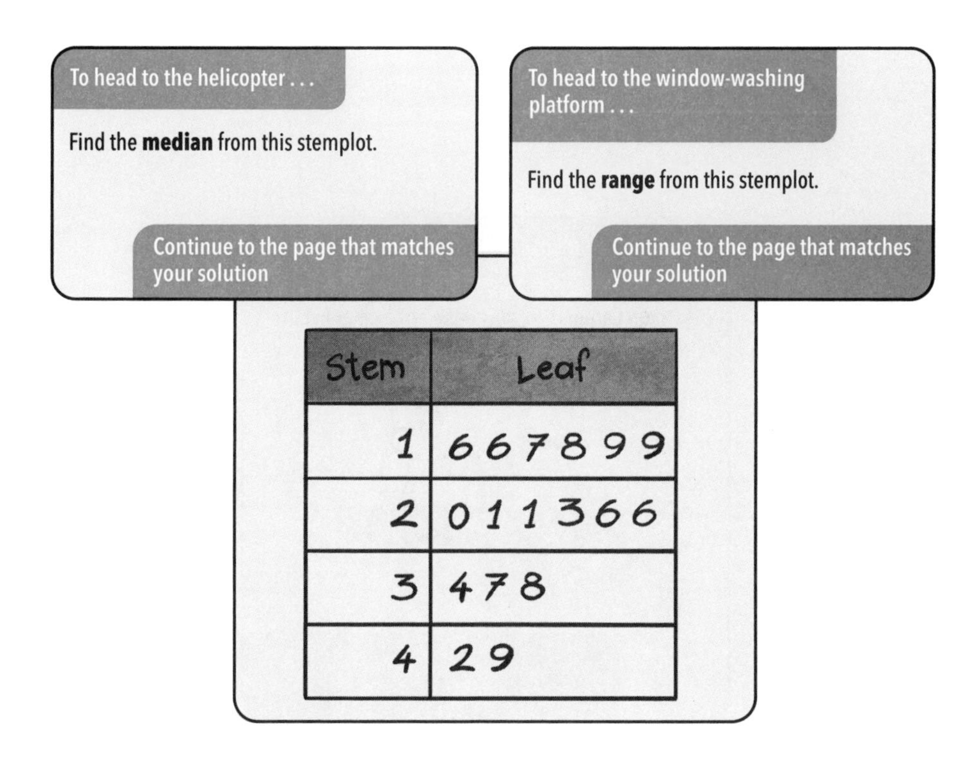

To head to the helicopter . . .

Find the **median** from this stemplot.

Continue to the page that matches your solution

To head to the window-washing platform . . .

Find the **range** from this stemplot.

Continue to the page that matches your solution

Stem	Leaf
1	6 6 7 8 9 9
2	0 1 1 3 6 6
3	4 7 8
4	2 9

See pages 55 and 60 of Adventurer's Advice Chapter 4 for help.

You dash down to the fifth floor just before the guards catch up with you, and run onto the landing through the stairway door.

The fifth floor is well lit, and it smells like a barn. You dash past some cubicle workspaces and into a large rectangular room.

One main table sits in the middle of the room, surrounded by dozens of caged animals. Some papers on the table are titled "Dregg Corporation Cosmetic Trials."

Cosmetic trials? Like makeup?

Sure enough, there are dozens of jars of face cream, tubes of mascara and lipstick, and jars of powder in tubs lined up against the back wall.

You open one of the cages and a magnificent pig struts out. It has curly amber hair in tight ringlets framing its face, rosy red lipstick, and long black eyelashes.

This is troubling, you think to yourself. More troubling still is the fact that the only door out of here is the one you came in through . . .

See page 52 of Adventurer's Advice Chapter 4 for help.

You should have arrived here from page 34

You weigh his offer for a moment. On the one hand, what Dregg and his cronies have been doing is clearly illegal. Their petty crime has been funding the unchecked growth of the Dregg Corporation. *Stealing is bad*.

On the other hand, if you get in on this scheme, you could get rich—quick. The cash you found in the wallet is small change compared to what you could make robbing museums and banks. And you won't have to run from Dregg and his goons anymore. *You could* be *a goon!*

You reach out your hand and meet his in a firm handshake. His mustache curls with the sly smile underneath. "Looks like we have a deal," he says dryly. "Let's head back to the tower and get you fitted for a henchman jumpsuit."

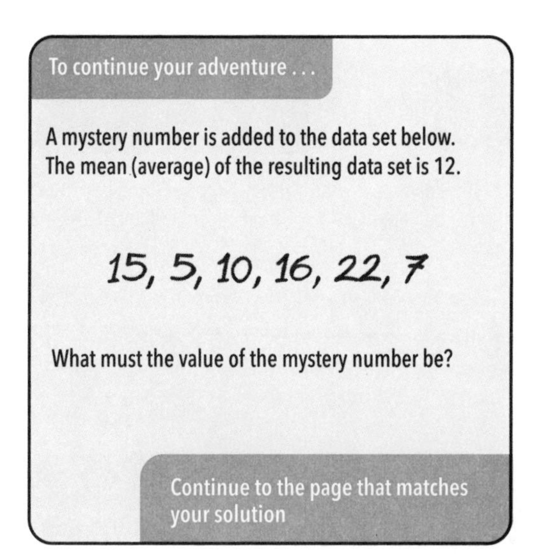

See page 58 of Adventurer's Advice Chapter 4 for help.

You should have arrived here from page 13, 18, 30, OR 43

The Egyptian Room has glass walls and high ceilings, and Egyptian artifacts surrounding a fountain filled with coins. The fountain isn't running now, but the water still makes the room feel pleasantly humid. There are some glass cases of smaller artifacts, and beyond them an impressive sarcophagus that was once the tomb for a mummy.

You can hear talking beyond the sarcophagus, and you are careful to step softly as you approach. You peer around the corner of the thing and see two figures. Facing you and sitting in a chair is a petite woman with long black hair. You recognize her right away. It's Doctor LaBella! Closer to you, and facing the doctor, is a tall figure holding a string with a pendant on the end. His features are cloaked in shadows, but you can hear his voice. "Stare at the amulet, Doctor, and repeat after me. 'Dregg is family. Dregg is power.'" As he recites this bizarre incantation, he is swinging the pendant back and forth. The doctor repeats his words back to him, almost as if in a trance.

She's hypnotized!

The man continues. "Very good, Doctor. Now, we won't be having any more disruptions to the plan, will we?" She shakes her head quietly. He puts the pendant into his pocket, satisfied with her answer, and walks past the doctor to a side door out of the Egyptian Room.

You must do something to get her out of here! You could splash Doctor LaBella's face with some water from the fountain, or you might be able to ask her some questions before the hypnosis wears off.

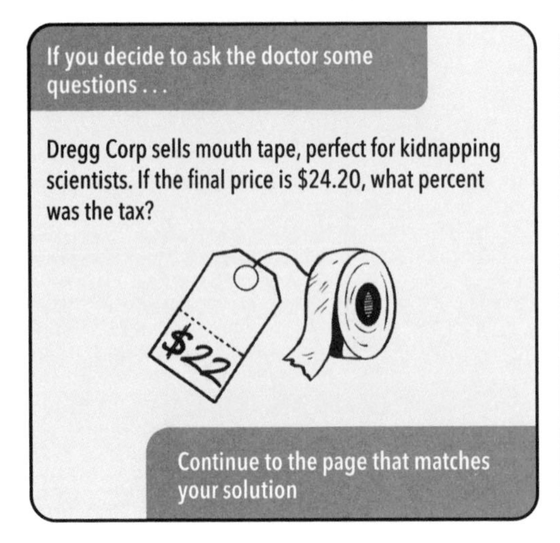

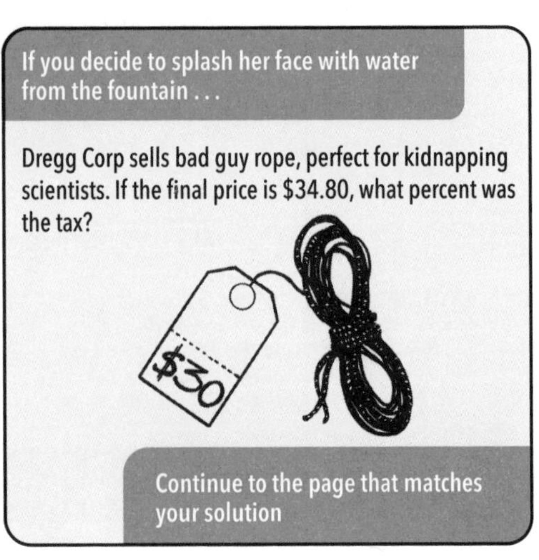

See page 53 of Adventurer's Advice Chapter 4 for help.

You should have arrived here from page 46

"Uh-oh," says Doctor LaBella, glancing behind the tank. Two Dregg security cars are tailing you with lights flashing. More importantly, the metal garage door up ahead is rolled down, blocking your escape.

"Let's see what these buttons do," says the doctor.

You press a big purple button on the dashboard, and a bright flash of light blasts from the long barrel of a weapon attached to the top of the tank. The flash blinds you for a moment, and when your eyes adjust, you can see that the gate has been frozen by some sort of . . . ice beam? *This thing has an ice beam!*

"Ooh. Do that again," says the doctor, as she pushes the gas pedal on the tank to the floor.

You zap the door again, and when the tank smashes into the gate, the frozen metal shatters into thousands of little pieces.

You are less than ten feet from escaping the tower, but the thousands of little metal shards from the frozen garage door bury the treads of the tank and clog up the rollers and wheels that make the machine move. Your tank stalls out, and you are stuck in place. More security cars are headed your way, and it's not looking like you will be escaping tonight.

The End

You should have arrived here from page 18

The Rare Art wing feels eerie and abandoned, and it's clear that Dregg has already been through here. The walls are painted a gray that's faded over time, but dark shadows of the paint's original color are left behind where paintings used to hang.

You see more droplets of blood on the pale carpeting of this wing. They get bigger and bigger, and lead to the closed door of a storage closet. In the drop of blood you see part of a print—maybe a hoof?—like it was stepped in. The door is locked, but you think you could force it open if you try.

Suddenly, you hear someone scream. It sounds like it's coming from the Egyptian Room. Maybe you should go there?

To force open the storage room door . . .

Find the mean (average) of the data set below.

41, 23, 35, 19, 47, 21

Continue to the page that matches your solution

To run to the Egyptian Room . . .

Find the mean (average) of the data set below.

35, 20, 17, 21, 43, 32

Continue to the page that matches your solution

See page 56 of Adventurer's Advice Chapter 4 for help.

You should have arrived here from page 30

The storage room is dark, and it goes back a ways. You can't see what's inside, but the trail of bloody dots continues into the darkness.

You take a few steps inside, and something drops from above, making a crashing sound all around you.

"We got 'em," says a gravelly voice as the lights turn on, revealing your fate.

You are trapped in a giant metal cage, surrounded by federal agents.

"You walked right into that! I can't *believe* you fell for the old 'trail of blood drops' trick!" one of the agents says to you through the metal bars of your cage. "You figured out how to rob this place, but you fell for the oldest move in the book!"

"This wasn't me!" you protest. "I'm on your side! I was after the thieves myself!"

But of course no one cares about your side of the story. You were caught red-handed, and nothing you can say at this point will clear your name.

The End

You should have arrived here from page 49

You dash through the tall glass doors into a large atrium filled with giant marble statues. *These must have been too heavy to steal.*

Your eyes scan the room for a way out. There are Greek figures chiseled from marble. There are Chinese terra-cotta warriors. All the statues in the atrium are arranged around a giant woolly mammoth skeleton that has been assembled in the center of the room, casting irregular shadows on the other statues.

There is a door past the skeleton, and it's the only door out of the room.

You sprint across the room toward the door, but a shadowy figure steps out from behind a stone pillar to block your path.

"About time we met face to face." The figure pulls back his hood, revealing icy blue eyes and pudgy, rosy cheeks, partially obscured by a bushy mustache. *Is it? It can't be.* The man speaks again. "My name is Levi Dregg."

To continue your adventure . . .

Find the median from the stemplot shown.

Stem	Leaf
1	9
2	2778
3	01124577
4	69
5	3
6	01
7	
8	5

Continue to the page that matches your solution

See page 55 of Adventurer's Advice Chapter 4 for help.

You should have arrived here from page 25

The roof of the museum is covered in solar panels, and you weave between them to reach the window-washing platform. It is suspended over the edge by two thick metal cables attached to a winch and pulley system. The platform itself is two feet wide and fifteen feet long, and there is a bucket and squeegee that one of the workers must have left after their shift.

You step onto the platform and it swings slightly under your weight.

A small, handheld control console on the platform has a few buttons on it, and you jam your thumb against the "down" arrow.

See ya, suckers!

Nothing happens.

You can see that the Dregg security guards have figured out the hatch, and a few are emerging onto the roof from the museum fire escape.

The platform isn't moving, and you hit the control console against your hand. Again, you push the "down" arrow. Again, nothing. *Uh-oh.*

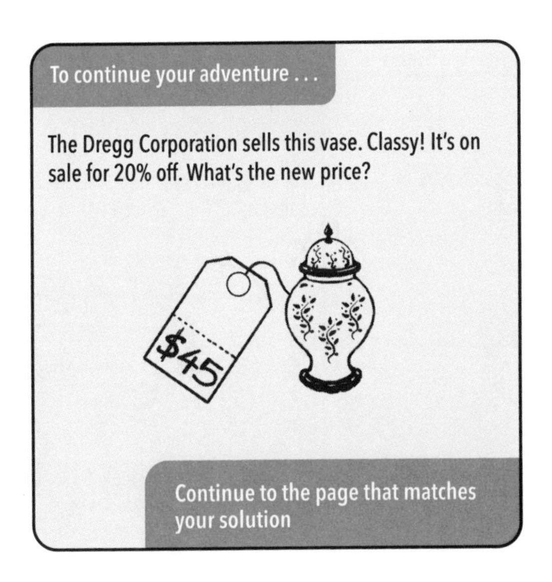

See page 52 of Adventurer's Advice Chapter 4 for help.

You should have arrived here from page 32

Levi Dregg? Is this really him? He looks exactly like the man you have seen on the news (and in hologram form in the Dregg Tower lobby). This real-life version is a little older and a little more disheveled. He is wearing tan work pants and a black, hooded jacket, zipped up to his chin.

"You're alive!" you exclaim with surprise. "And look at this mess you made! *Why?*"

"Our second-quarter projections this year were too low, and we needed some quick money to keep our investors happy," he explains. "Doctor LaBella cracked the portal technology last year, and I saw a perfect opportunity to increase our profit margins and continue expanding. The banks and museums Dregg Corp stole from—that's just business," he continues flatly.

"What I didn't count on was you sticking your nose where it didn't belong." The tone of his voice is pointed now, and he takes a step toward you. "So here's the deal, kid. You can make some money and help out with this operation. I'll even throw you some stock options. It's either that or you're against us. And you won't like what happens if you're against us. Just ask Doctor LaBella. She developed these technologies but she didn't like where they led. She didn't share my vision, and it cost her everything." He shakes his head before turning back to you.

You can take Levi Dregg up on his offer and join him, or you can make a run for the door past him.

See page 52 of Adventurer's Advice Chapter 4 for help.

You should have arrived here from page 11

There must be six or seven security guards standing in the goopy agar of the sixth floor, but they don't seem surprised or bothered by the goo. You need to stop them and escape.

You glance around the lab, looking for a way out. There is a door against the back wall, just beyond a huge 55-gallon drum labeled "Dregg Cell Nutrients: For Rapid Colony Growth." The drum doesn't have a lid, and you can see a small measuring spoon resting in the orange nutrient powder.

You pour a single spoonful of the stuff onto the blobby cells pulsating on the floor. They immediately begin to grow and divide, pulsing happily with the meal you have provided. You turn back to the barrel and, using your whole weight, tip it over, spilling orange nutrient powder all over the huge single cells.

Almost instantaneously, they begin to pulsate with this extreme influx of food. In a matter of moments, the guards are completely hidden from view behind a wall of bulbous bacteria. The entire lab is filling rapidly . . .

To continue your adventure . . .

A mystery number is added to the data set below. The mean (average) of the resulting data set is 18. What must the value of the mystery number be?

16, 12, 23, 15, 9, 14

Continue to the page that matches your solution

See page 58 of Adventurer's Advice Chapter 4 for help.

You should have arrived here from page 33

The Dregg guards have spotted you from across the roof and are fast approaching. The window-washing platform isn't moving, and it doesn't seem to have power.

You step off the platform back to the roof, and look around. The winch system is connected to a big blue generator sitting on the roof beside it. You grab the pull cord and give it a swift pull. The generator rumbles, and after a second pull it jumps to life.

You hop back onto the platform, and this time, when you press the "down" arrow on the handheld control, the platform begins to descend.

To continue your adventure . . .

A mystery number is added to the data set below. The mean (average) of the resulting data set is 19. What must the value of the mystery number be?

22, 6, 15, 32, 11, 43

Continue to the page that matches your solution

See page 58 of Adventurer's Advice Chapter 4 for help.

You should have arrived here from page 35

The cells between you and the guards are pulsating and rumbling as they ravenously devour the nutrient powder. *What are they going to eat when they finish this food?*

You hear screams from the security guards on the other side of this giant experiment gone wrong. *You may have found the answer to your question.*

You turn to the door, but the rapidly replicating cells have blocked your exit. They have you surrounded, and they continue to grow and divide until you are buried beneath a mountain of mitochondria and membranes. The cells are hungry.

The End

You should have arrived here from page 17

You hop into the front seat of the hovercraft next to Doctor LaBella. She pulls down the visor and a small set of keys drops into her lap. "Jerry always leaves the keys in this one."

As she turns the key in the ignition, the engine rumbles to life. She takes a wide turn out of the parking spot, knocking over a helicopter being stored in the garage. "Oops. I hope that's not expensive," she jokes.

The air propelling the hovercraft is deafening in the confines of the garage, but it only takes a minute before the massive vehicle reaches the exit and the dark, cool night outside.

She glances over her shoulder. "Looks like we got company. Hang on!"

To continue your adventure . . .

A mystery number is added to the data set below. The mean (average) of the resulting data set is 27. What must the value of the mystery number be?

38, 13, 41, 26, 13, 35

Continue to the page that matches your solution

See page 58 of Adventurer's Advice Chapter 4 for help.

Go back to the previous page and check your work. You should not have arrived here.

You should have arrived here from page 1

You don't want to open either of the portals that you just closed, but your curiosity takes over. There is a third control panel. Without thinking it through, you slide the purple Dregg card into the access slot. Nothing happens.

You try Doctor LaBella's blue ID card. Almost immediately, the third control panel jumps to life with a flurry of blinking lights and humming machinery.

Across the lab, a small circle of glowing blue plasma begins to build, framed by the large metal structure. The lights around you flicker as the portal pulls more and more energy and the spinning blue ether grows. The ground shakes, and the lights flicker more and more violently until they go out entirely.

This explains the energy surges that you saw in the lobby, you think to yourself. These portals need a TON of electricity to get going.

The swirling blue plasma of the portal fills the entire structure now, and as the lights around the lab slowly blink back to life, you peer inside. Where does this one lead? You see a woman in the center of the space on the other side. She looks right at you.

"Can you help me?" she asks you. "I think you're my only hope."

It's Doctor LaBella!

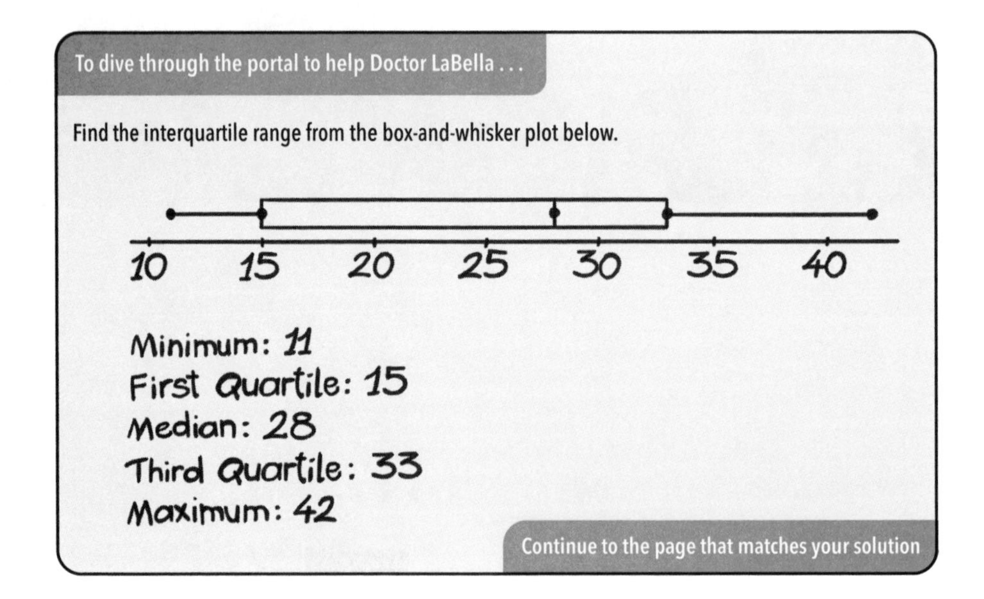

See page 59 of Adventurer's Advice Chapter 4 for help.

You should have arrived here from page 10

The doctor has fully broken out of her hypnotic state, and she is fast on her feet. You race step for step with her down the museum hallway to the portal. You leap through first, and she is only a half-step behind you. You both land with a thud back on the eleventh floor of Dregg Tower.

You power down the portal to ensure that nobody from the museum will catch up with you. Doctor LaBella shoves her own ID and the two other activation cards for the portals into the front pocket of her lab coat. "Let's see them get the portals running without these!" she exclaims.

An alarm is ringing in the distance. "They have definitely sent security to find us, so we can't take the elevator or the stairs, but I have an idea!"

You follow her to a blackboard against the back wall, and it slides to reveal a secret elevator. "Dregg had this installed to smuggle out all the stuff they have been stealing with the portals. It only stops on two levels: the loading dock in the parking garage, and the second floor where the bigwigs have their offices."

You step inside with the doctor and the doors slide shut behind you.

If you tell Doctor LaBella you think you should head to the parking garage . . .

Use the two-way frequency table below. What percent of all those surveyed are warehouse workers who work the day shift?

Continue to the page that matches your solution

If you insist on going to the second floor . . .

Use the two-way frequency table below. What percent of night shift employees are security guards ?

Continue to the page that matches your solution

	Day Shift	Night Shift	Total
Security Guards	18	4	22
Warehouse Workers	12	16	28
Total	30	20	50

See page 62 of Adventurer's Advice Chapter 4 for help.

You should have arrived here from page 19 OR page 22

Doctor Donda LaBella leads you to the left at the second fork in the maze. A bright red light shines from the end of the makeshift hallway in front of you. Doctor LaBella puts her arm out to block your path. "Hang on." You both freeze in your tracks.

"The maze administers shocks when the lights are red. It tests the memory of the rats. Don't. Move. An. Inch."

The two of you stand paralyzed, staring at the red light down the hallway.

The light switches to green, and you continue cautiously to a wide section of the corridor. In the center, there is a big red button on a pedestal at belly height.

You stand across the button from Doctor LaBella. Her brow furrows. "What do you think? Should we push it?"

Push the button . . .

A mystery number is added to the data set below. The mean (average) of the resulting data set is 15. What must the value of the mystery number be?

13, 9, 26,
14, 23, 17

Continue to the page that matches your solution

Don't push the button . . .

A mystery number is added to the data set below. The mean (average) of the resulting data set is 11. What must the value of the mystery number be?

10, 18, 15,
8, 10, 14

Continue to the page that matches your solution

See page 58 of Adventurer's Advice Chapter 4 for help.

You should have arrived here from page 13

"Have we met before?" you ask Mr. Lucky. "You look so familiar."

Mr. Lucky smiles but doesn't say anything.

You'd love his help, but what can you say? *I'm following a crime ring of lab animals and robots controlled by Levi Dregg as they travel through blue portals that transcend space and time?* That all sounds ridiculous.

You blurt out something else instead, "You, uh, work for the museum?"

Nice save.

Again, he stays perfectly silent, but points down the hall toward the Egyptian Room. His eyes seem to convey an important message.

Even though he doesn't say a word, you know exactly where you will find Donda LaBella.

To go to the Egyptian Room, turn to page 28.

You should have arrived here from page 34

Join Levi Dregg? Become some sort of sci-fi criminal? *No thank you.*

You stall for a minute by running your hand through your hair as you eye the door in the corner. "How much do you think I can make?" you ask coyly.

The corner of his mouth curls in a crooked smile, and as he starts to answer, you inch slowly to your right, looking for an opening.

Two of the guards from the portal enter behind you, and they draw Levi Dregg's attention for a moment. That moment is all you need, and you bolt toward the door. Levi Dregg is past his prime, and you are able to outrun him easily. You push the door open and then quickly shut and lock it behind you.

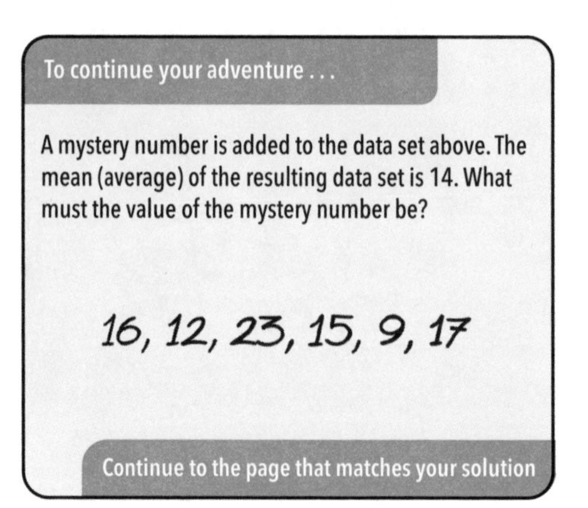

See page 58 of Adventurer's Advice Chapter 4 for help.

You should have arrived here from page 19 OR page 22

Doctor Donda LaBella leads you to the right at the second fork in the maze. The floor in front of you is made from giant hexagonal tiles. The tiles must be made of glass or plastic, and the one in front of you is lit up from a light below.

Doctor LaBella crouches down to inspect the tile. "These have weight sensors on them," she says, rubbing her chin and thinking deeply about something.

Finally she says, "We can only stand on the lit-up ones. The maze administers a shock if you screw up, and the shocks are calibrated for a 400-pound rat. So, follow my lead unless you want to find out what that feels like."

Doctor LaBella stands on the first tile, and you join her. Your weight immediately triggers the weight sensor, and a tile two away lights up. You both leap across the tile in between and land on this new tile, triggering the next tile in the sequence.

One tile at a time, you carefully navigate across this section of maze until the floor returns to normal, and you round a corner.

You continue cautiously to a wide section of the corridor, but you aren't safe yet. In the center, there is a big red button on a pedestal at about belly height.

You stand across it from Doctor LaBella. Her brow furrows. "What do you think? Should we push it?"

Push the button . . .

A mystery number is added to the data set below. The mean (average) of the resulting data set is 15. What must the value of the mystery number be?

13, 9, 26,
14, 23, 17

Continue to the page that matches your solution

Don't push the button . . .

A mystery number is added to the data set below. The mean (average) of the resulting data set is 11. What must the value of the mystery number be?

10, 18, 15,
8, 10, 14

Continue to the page that matches your solution

See page 58 of Adventurer's Advice Chapter 4 for help.

You should have arrived here from page 17

You hop in the front seat of the tank next to Doctor LaBella. She reaches under the steering wheel and pulls out a fistful of colorful wires. “Red to black, and . . .”

As she twists two of the wires together, the engine rumbles to life. Doctor LaBella takes a wide turn out of the parking spot, the massive tank treads running over three expensive-looking motorcycles. “Oops,” she says. “They can take that off my last paycheck.”

The tank rumbles up the ramp, toward the exit.

To continue your adventure . . .

A mystery number is added to the data set below. The mean (average) of the entire data set is 25. What must the value of the mystery number be?

28, 31, 20, 19, 22, 26

Continue to the page that matches your solution

See page 58 of Adventurer's Advice Chapter 4 for help.

Go back to the previous page and check your work. You should not have arrived here.

You should have arrived here from page 21

The museum is across town from Dregg Tower. There aren't many tall buildings here for you to try to reach with the grappling hook.

The west side of the museum abuts a park, and there is a tall maple tree that is maybe twenty yards from your perch atop the museum roof.

If you can get the grappling hook around a sturdy tree branch, you might be able to escape.

A noise behind you makes you turn, and you can see a security guard clambering through the hatch onto the roof.

Now or never. You aim the grappling hook at the tree and pull the trigger.

The barbed grappling hook sails through the air and lands among the foliage of the maple tree below you in the park. You can't see if it hooked onto anything, but you tug on it as hard as you can, and it doesn't budge.

Now you need to secure the other end of the line. You wind the steel cable around a stone gargoyle perched on the edge of the roof and test your line. *It seems secure, but there is only one way to know for sure.*

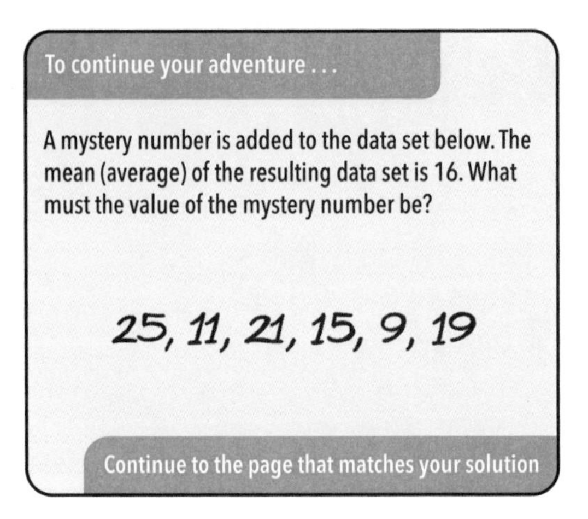

See page 58 of Adventurer's Advice Chapter 4 for help.

You should have arrived here from page 16

You run past the blue plasma of the portal and find yourself in the museum's huge front exhibit hall.

Tall windows let in light from the moon, and you can see more carnage from the museum robbery. There are open spots on the walls where paintings should be, and there are shards of glass on the floor from broken display cases.

You can hear footsteps coming from the portal. The Dregg security guards can't be far behind you. You look around frantically for an exit.

There are metal rungs leading to the roof against the wall to your left. It looks to be some sort of fire escape. There is also a large glass door that leads to a statue garden outside. It's been propped open. You hesitate. The fire escape is dangerous, but the propped door suggests someone else came this way.

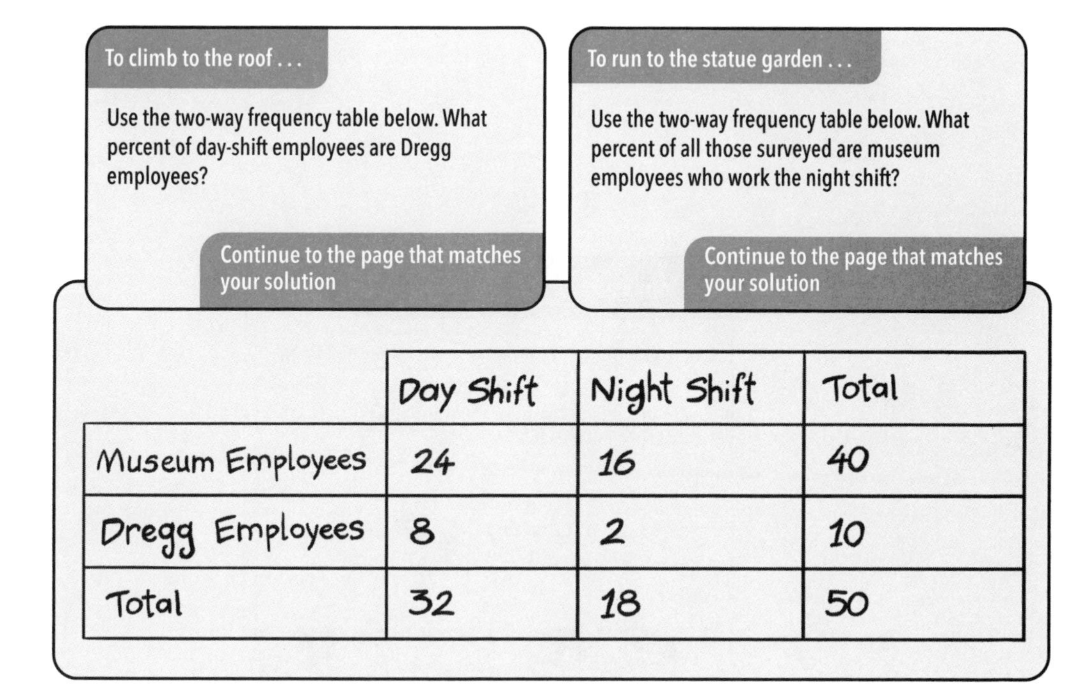

To climb to the roof . . .

Use the two-way frequency table below. What percent of day-shift employees are Dregg employees?

Continue to the page that matches your solution

To run to the statue garden . . .

Use the two-way frequency table below. What percent of all those surveyed are museum employees who work the night shift?

Continue to the page that matches your solution

	Day Shift	Night Shift	Total
Museum Employees	24	16	40
Dregg Employees	8	2	10
Total	32	18	50

See page 62 of Adventurer's Advice Chapter 4 for help.

You should have arrived here from page 26

You are trapped in this weird lab with a bunch of farm animals wearing makeup. Dregg security is no doubt a few minutes from finding you.

You open another cage, and a white sheep with a tight perm, wearing thick, caked-on foundation, saunters into the lab space.

You run around the lab releasing all of the animals. The room is filled with goats, sheep, pigs, dogs, chickens, and raccoons, all wearing a preposterous amount of blush, mascara, and eye shadow. You open the door to the room, and they storm out. Captivity and cheap bronzer have driven them into a frenzy, and now that they have been released, they are bursting with frantic energy.

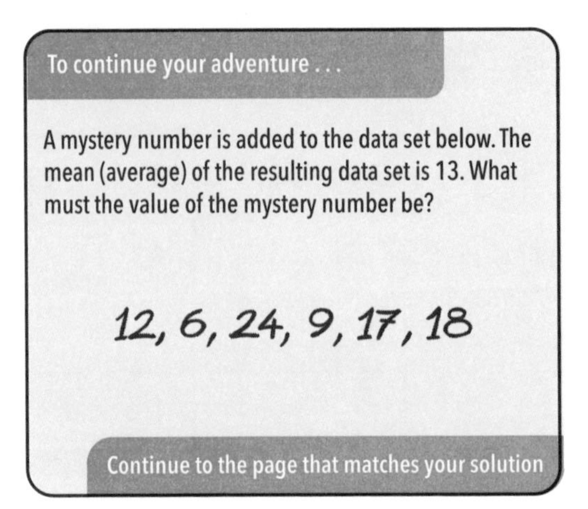

See page 58 of Adventurer's Advice Chapter 4 for help.

Probability (spinner) (page 1)

This problem asks us to find the probability that the spinner lands on a 1 or a 3.

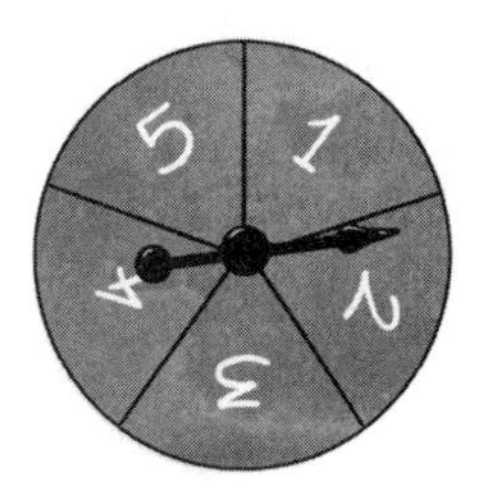

Space 1 or Space 3 → 2 / 5 possible spaces on spinner → 5

$$\frac{2}{5}$$

STEP ONE: Find the probability as a fraction. There are 5 even spaces on the spinner, and assuming all the spaces are equally likely, we have a 2/5 chance of hitting 1 or 3.

STEP TWO: Find the solution as a percent. We can do this a couple ways. The easiest way is to turn that fraction into a decimal, and then make that decimal into a percentage. 2/5 = 0.40. We have found our percent as a decimal, and to turn that decimal into a percent, we need to multiply it by 100. Remember that "percent" means "per 100."

$$\underbrace{\frac{2}{5}}_{\text{fraction}} = \underbrace{0.40}_{\text{decimal}} = \underbrace{40\%}_{\text{percent}}$$

We know that 0.40 represents "forty hundredths," "40 per 100," or "40 percent." Continue to that page!

ANSWER KEY FOR THIS TYPE OF PROBLEM:

Page 1: Continue to page 40

Finding a new price (pages 11, 17, 19, 21, 22, 26, 33, and 34)

This problem gives us a $40 tennis racket and asks us to find the new price after a 12.5% discount.

12.5% → 0.125
percent decimal
$40.00 × 0.125 = $5.00

STEP ONE: Find the amount of the discount. How much money is 12.5%? It helps to remember that *percent* means "per 100." How do we write that "per 100" with decimals? The second place after the decimal point is the hundredths place, so 0.01 would be 1 per 100. 12.5 percent is 12.5 per 100 and can be written as 0.125. If we multiply that discount by our price, we find that we will save $5.

STEP TWO: Find the new price. We are saving $5 off of a $40 item, so to find the new price, we need to subtract: 40 − 5 = 35. The new price of the racket is $35, so continue to page 35.

$40.00 − $5.00 = $35.00
price discount new price

AUTHOR'S NOTE: Another great way to solve this is to think of what percent of the racket we will actually be paying for. If we are *saving* 12.5%, that leaves 87.5% of the racket that we *still need to pay for*. $40 times 0.875 will also give us a price of $35.

ANSWER KEY FOR THIS TYPE OF PROBLEM:

Page 11: Continue to page 35
Page 17 (left): Continue to page 38
Page 17 (right): Continue to page 46
Page 19 (left): Continue to page 42
Page 19 (right): Continue to page 45
Page 21: Continue to page 48
Page 22 (left): Continue to page 42
Page 22 (right): Continue to page 45
Page 26: Continue to page 50
Page 33: Continue to page 36
Page 34 (left): Continue to page 44
Page 34 (right): Continue to page 27

Finding a missing percent (page 28)

This problem deals with percents, and it asks how much tax we paid on a roll of tape if the original price was $22 and the price after tax is $24.20.

$24.20 - $22.00 = $2.20
final price — price before tax — amount of tax

STEP ONE: Find the amount the price changed. In this problem, our price increased from $22 to $24.20. The difference between prices is $2.20.

STEP TWO: Represent the percent change as a fraction. Our $22 roll of tape increased in price by $2.20. We paid $2.20 in tax per $22. We can express this as a fraction: 2.20/22.

$2.20 ← tax
$22.00 ← price before tax

$2.20/$22.00 → 0.10 → 10%
fraction — decimal — percent

STEP THREE: Find the tax as a percent. We can do this a couple ways. The easiest way is to turn that fraction into a decimal, and then make that decimal into a percentage. 2.20/22 = 0.10. We have found our percent as a decimal, and to turn that decimal into a percent, we need to multiply it by 100.

Remember that "percent" means "per 100." We know that 0.10 represents "ten hundredths," "10 per 100," or "10 percent." Continue to that page!

AUTHOR'S NOTE: We could also solve Step Three by setting up a proportion. Remember that "percent" means "per 100." If we find the missing numerator that makes x/100 equal to 2.20/22, we can also find our missing percentage. Check back to Chapter 1, page 58 of the Chapter 1 Adventurer's Advice, for a reminder on how to solve this problem with that strategy!

ANSWER KEY FOR THIS TYPE OF PROBLEM:

Page 28 (left): Continue to page 10
Page 28 (right): Continue to page 16

Median (pages 10 and 16)

This question is asking us to find the median of this set of numbers. Median is a *measure of center.* Measures of center (commonly median or mean/average) tell us about the middle of the group of numbers. Where are these numbers clumped up at?

62, 37, 56, 40, 15, 22, 65, 42, 25, 47

Median specifically wants us to identify the middle-est value from our data set.

In order:
15, 22, 25, 37, 40, 42, 47, 56, 62, 65

STEP ONE: Put the data in order from smallest to largest. It will be easier to identify the middle data point if we are looking at the data in order.

STEP TWO: Cross off pairs: 1 minimum and 1 maximum. Cross off the smallest and the biggest value. Repeat until you have only one value left. That number is your median.

Cross off in pairs
~~15~~, 22, 25, 37, 40, 42, 47, 56, 62, ~~65~~
~~15~~, ~~22~~, 25, 37, 40, 42, 47, 56, ~~62~~, ~~65~~
etc...
~~15~~, ~~22~~, ~~25~~, ~~37~~, 40, 42, ~~47~~, ~~56~~, ~~62~~, ~~65~~

2 data points left!
40, 42 — median in between!

STEP THREE (IF NECESSARY): Find the average/mean of the two central data points. There are an even number of data points in this set, which means that we don't have *one* most central data point. We have *two.* For our data set, 40 and 42 are the central data points. If we find the number between those two points, we have our median for this data set! The number between 42 and 40 is 41. 41 is our median, so continue to that page!

ANSWER KEY FOR THIS TYPE OF PROBLEM:

Page 10: Continue to page 41
Page 16 (left): Continue to page 7
Page 16 (right): Continue to page 49

Median from a stemplot (pages 15, 20, 25, and 32)

This data representation is called a stemplot or sometimes a stem-and-leaf plot. It uses discrete values (no decimals or fractions), and it can a good way to visualize the shape of a set of data. The left column labeled "stem" represents our 10s, so this stemplot includes numbers in the tens, twenties, thirties, and forties. The right column, labeled "leaf," represents our ones, and each value in the leaf column represents a different number. A stem of 1 with a leaf of 6 together make 16. This stemplot includes 16, 16, 17, 18, 19, 19, 20, 21 . . .

Stem	Leaf
1	667899
2	011366
3	478
4	29

(Note that "0" actually represents 20 because the 0 is with a stem of 2.)

This question is asking us to find the median of this set of numbers. Median is a *measure of center.* Measures of center (commonly median or mean/average) tell us about the middle of the group of numbers. Where are these numbers clumped up at?

Median specifically wants us to identify the middle-est value from our data set.

Stem	Leaf
1	667899
2	011366
3	478
4	29

16, 16, 17, 18, 19, 19, 20, 21, 21,
23, 26, 26, 34, 37, 38, 42, 49

STEP ONE: List out the values from the stemplot. If you feel comfortable with stemplots, you can skip this step because the values are already in order, but it makes it easier to see all the values if we list them out in order.

STEP TWO: Cross off pairs: 1 minimum and 1 maximum. Cross off the smallest and the biggest value. Repeat until you have only one value left. That number, 21, is your median.

~~16~~, 16, 17, 18, 19, 19, 20, 21, 21,
23, 26, 26, 34, 37, 38, 42, ~~49~~

~~16~~, ~~16~~, 17, 18, 19, 19, 20, 21, 21,
23, 26, 26, 34, 37, 38, ~~42~~, ~~49~~

Continue until...

~~16~~, ~~16~~, ~~17~~, ~~18~~, ~~19~~, ~~19~~, ~~20~~, ~~21~~, (21),
~~23~~, ~~26~~, ~~26~~, ~~34~~, ~~37~~, ~~38~~, ~~42~~, ~~49~~

ANSWER KEY FOR THIS TYPE OF PROBLEM:

Page 15 (right): Continue to page 11
Page 20 (left): Continue to page 19
Page 25 (left): Continue to page 21
Page 32: Continue to page 34

Mean (pages 13 and 30)

This question is asking us to find the mean of this set of numbers. Mean is a *measure of center.* Measures of center (commonly mean/average or median) tell us about the middle of the group of numbers. Median asks for the middle number while mean asks how much value each data point would have if they were all equal.

38, 51, 19, 44, 60, 46

$38 + 51 + 19 + 44 + 60 + 46 = 258$

STEP ONE: Find the total. Add up all the numbers in the data set. The total for this data set is 258.

STEP TWO: Divide by the number of data points in the set. We have six data points, so we will evenly divide the 258 from Step One into six even groups. The average for this data set is 43, so continue to that page!

$\frac{258}{6} = 43$

ANSWER KEY FOR THIS TYPE OF PROBLEM:

Page 13 (left): Continue to page 43
Page 13 (right): Continue to page 28
Page 30 (left): Continue to page 31
Page 30 (right): Continue to page 28

Missing values on a two-way frequency table (page 18)

This is called a two-way frequency table, and these tables are a good way to collect data when we are measuring more than one characteristic. This table is looking at a group of Dregg employees. Each cell on the table shows a different group of those employees. The 17 in the table tells us that there are 17 researchers that have worked at Dregg for 0 to 5 years. The 40 tells us that there are 40 total developers. We need to find the numbers for the missing cells, and to do that we need to use our "total" column and row.

	0-5 years	6-10 years	11-15 years	Total
Researchers	17			
Developers	13		15	40
Total		42		100

	0-5 years	6-10 years	11-15 years	Total
Researchers	17			
Developers	13		15	40
Total	30	42		100

STEP ONE: Look for a value we can find. We know how many researchers have worked at Dregg for 0 to 5 years (17) and how many developers have worked at Dregg for 0 to 5 years (13). That means that there are a total of 30 employees who have worked for 0 to 5 years in this data set. We can put 30 in that cell, and it will help us find other values in the table.

STEP TWO: Repeat Step One until the table is completed. There are 30 employees who have worked for 0 to 5 years and 42 who have worked for 6 to 10 years. We also know that there are 100 total employees in the data set, so that means that there are 28 employees left who have worked for 11 to 15 years, and we can enter that value in the corresponding cell.

	0-5 years	6-10 years	11-15 years	Total
Researchers	17	30	13	60
Developers	13	12	15	40
Total	30	42	28	100

	0-5 years	6-10 years	11-15 years	Total
Researchers	17	30	(13)	60
Developers	13	12	15	40
Total	30	42	28	100

If we want to find the number of researchers who have worked at Dregg for 11 to 15 years, find the cell that corresponds with that "researchers" and "11–15 years." There are 13 employees who fit both criteria, so continue to page 13!

ANSWER KEY FOR THIS TYPE OF PROBLEM:

Page 18 (left): Continue to page 30
Page 18 (center): Continue to page 13
Page 18 (right): Continue to page 28

Given mean, missing value (pages 27, 35, 36, 38, 42, 44, 45, 46, 48, and 50)

This problem asks us to look at mean (aka average) a little differently. It gives us *most* of the numbers: 15, 5, 10, 16, 22, and 7. What it doesn't give us is the missing number. To help us find that missing number, we know that the mean for all the numbers together will be 12.

15, 5, 10, 16, 22, 7

Mean Calculation:

$$\frac{15+5+10+16+22+7+x}{7}$$

$$\frac{15+5+10+16+22+7+x}{7}=12$$

STEP ONE: Set up an equation. In this problem, let's pick the variable x to represent the number we are missing. Check out page 56 of this chapter's Adventurer's Advice if you forgot how to find the mean, but basically we add all the values together and divide by *how many* values we have. If we count x, we have 7 values, so our expression would say (15 + 5 + 10 + 16 + 22 + 7 + x)/7.

The problem also tells us that the mean with our x value is 12, so we can set our big fraction equal to 12.

STEP TWO: Solve the equation that we just made. Combine the like terms in the numerator to get (75 + x)/7 = 12. Use inverse operations to get rid of the "divided by 7" first, and then subtract 75 on both sides. We find that the missing value has to be 9, so continue to page 9!

$$\frac{15+5+10+16+22+7+x}{7}=12$$

$$(\cdot 7)\frac{75+x}{7}=12(\cdot 7)$$

$$75+x=84$$

$$-75 \quad -75$$

$$x=9$$

ANSWER KEY FOR THIS TYPE OF PROBLEM:

Page 27: Continue to page 9
Page 35: Continue to page 37
Page 36: Continue to page 4
Page 38: Continue on page 23
Page 42 (left): Continue to page 3
Page 42 (right): Continue to page 2
Page 44: Continue to page 6
Page 45 (left): Continue to page 3
Page 45 (right): Continue to page 2
Page 46: Continue to page 29
Page 48: Continue to page 12
Page 50: Continue to page 5

Interquartile Range (IQR) (page 40)

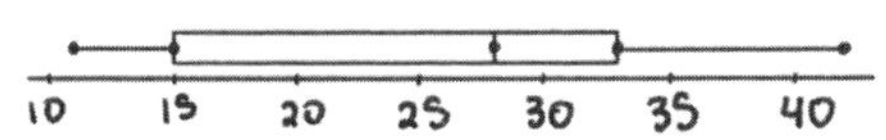

This data representation is called a boxplot, or sometimes a box-and-whisker plot. It can be a great way to visualize the way the data is distributed, and it shows a number of important features of our data set. The center of the box is located at the median, and the end of each whisker shows us the minimum and maximum.

A boxplot also divides up the data into four equal sections at values called "quartiles" The median is the middle quartile and it splits the data set in half. Each half is again split with the lower and upper quartiles. The lower quartile is sometimes called the "first quartile" or "Q1." Similarly, the upper quartile can also be called the "third quartile" or "Q3." If our data set has 20 data points, the quartiles would divide our data into four sections with 5 data points each. Each quarter of the data is easy to see on a boxplot as each of the two whiskers and the two halves of the box contain one fourth of the data points.

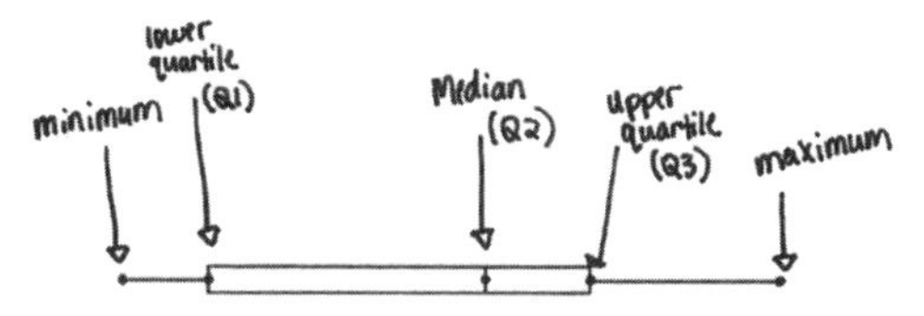

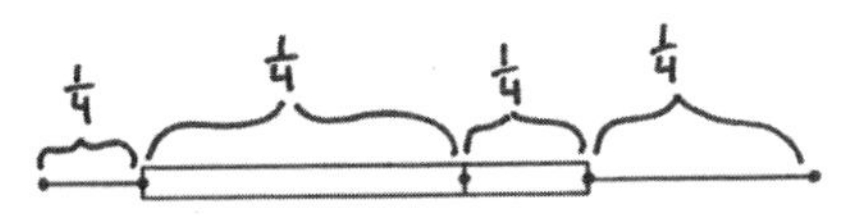

Quartiles are convenient, because we can easily see where the "top half" of the data are. In the example above, the top half would be between the median (28) and the largest value (42). We can also find the half of the data that is most central. The middle half of the data in the example above would be between (15) and (33). The *interquartile range* (IQR) is a measure of how big that "middle half" of the data is. It's basically the range of the box, from Q1 to Q3. This measurement in particular is important, because it will omit any outliers below or above the rest of the data points and give us a measure of how spread out our data is with any outliers removed.

33 - 15 = 18

STEP ONE: Identify the values of the lower quartile and the upper quartile. These points are at the edges of the box in the boxplot. The lower quartile in this boxplot is at 15 and the upper quartile is at 33.

STEP TWO: Find the difference. The lower quartile is 15, and the upper quartile is 33. To find the interquartile range, subtract: 33 − 15 = 18. That means that the middle half of our data points are spread across the 18-size interval from 15 up to 33, so continue to page 18!

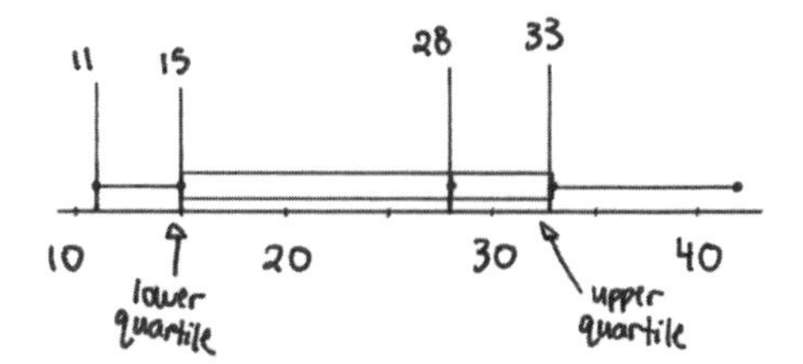

ANSWER KEY FOR THIS TYPE OF PROBLEM:
Page 40: Continue to page 18

Range from a stemplot (pages 15, 20, 24, and 25)

This data representation is called a stemplot or sometimes a stem-and-leaf plot. It uses discrete values (no decimals or fractions) and it can be a good way to visualize the shape of a set of data. The left column labeled "stem" represents our 10s, so this stemplot includes numbers in the tens, twenties, thirties, and forties. The right column, labeled "leaf," represents our ones, and each value in the leaf column represents a different number. A stem of 1 with a leaf of 6 together make 16. This stemplot includes 16, 16, 17, 18, 19, 19, 20, 21 . . .

Stem	Leaf
1	667899
2	011366
3	478
4	29

(Note that "0" actually represents 20 because the 0 is with a stem of 2.)

This question is asking us to find the range from this set of data. Range is a "measure of spread," which means it tells us about how spread apart or bunched together a group of numbers is. This question is asking us to find how big the *range* of numbers is.

Stem	Leaf
1	667899
2	011366
3	478
4	29

Smallest: 16
largest: 49

STEP ONE: Identify the smallest and largest data points. For this data, our smallest value is 16 and our largest value is 49.

STEP TWO: Find the difference. The smallest value is 7 and the largest is 15. To find the range, subtract: 49 – 16 = 33. That means that all our data points are spread across the 33-size interval from 16 up to 49, so continue to page 33!

49-16=33

ANSWER KEY FOR THIS TYPE OF PROBLEM:

Page 15 (left): Continue to page 26
Page 20 (right): Continue to page 22
Page 24: Continue to page 17
Page 25 (right): Continue to page 33

Range from a dotplot (pages 2 and 3)

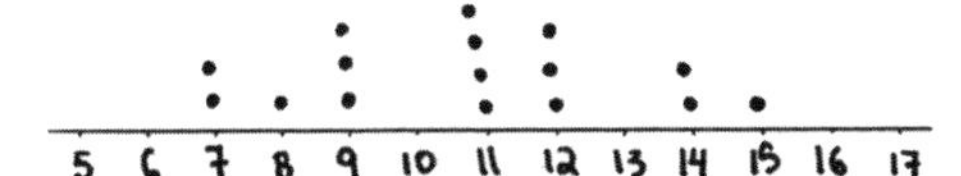

This data representation is called a dotplot. It uses discrete values (no decimals or fractions) and it can be a good way to visualize the shape of a set of data. There are two dots above 7, which indicate two 7s in this data: 7, 7, 8, 9, 9, 9, 11, 11, 11, 11, etc.

This question is asking us to find the range from this set of data. Range is a "measure of spread," which means it tells us about how spread apart or bunched together a group of numbers is. This question is asking us to find how big the range of numbers is.

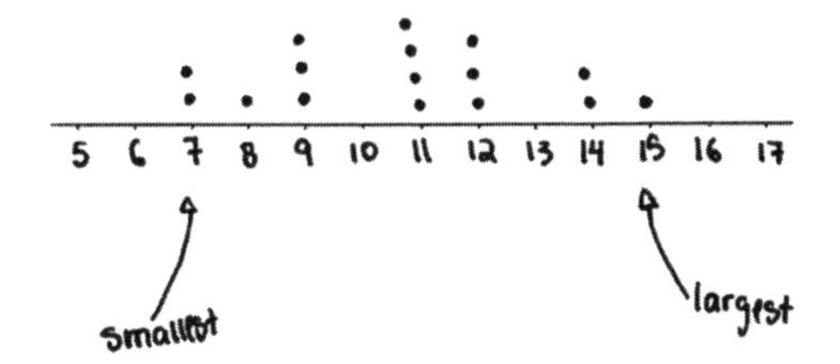

STEP ONE: Identify the smallest and largest data points. For this data, our smallest value is 7 and our largest value is 15.

STEP TWO: Find the difference. The smallest value is 7 and the largest is 15. To find the range, subtract: $15 - 7 = 8$. That means that all our data points are spread across the 8-sized interval from 7 up to 15, so continue to page 8!

$15 - 7 = 8$

ANSWER KEY FOR THIS TYPE OF PROBLEM:

Page 2: Continue to page 8
Page 3: Continue to page 14

Percent from a two-way frequency table (pages 7, 41, and 49)

What percent of the stolen jewels are sparkly?

This question asks us to find what percent of the stolen jewels are sparkly. How do we find that percentage? How do we even know which cells from this table are important?

	Sparkly	Shiny	Total
Jewels	9	51	60
Gems	14	16	30
Total	23	67	90

	Sparkly	Shiny	Total
Jewels	9	51	(60)
Gems	14	16	30
Total	23	67	90

STEP ONE: Identify the correct total. *This* question is only asking us about the stolen jewels. (We could add sparkly gems, or take them away, and that won't change our answer one bit!) On the right side of the table, we can see that there are 60 jewels, and for this problem that is our total.

STEP TWO: Answer the question as a fraction. Of the 60 jewels that we identified in Step One, only 9 are in the "sparkly" column. That means that 9 jewels *out of 60* are sparkly, or 9/60.

$\frac{9}{60}$ ← sparkly jewels / ← total jewels

$\frac{9}{60} \rightarrow 0.15 \rightarrow 15\%$

fraction decimal percent

STEP THREE: Find your solution as a percent. We can do this a couple ways. The easiest way is to turn that fraction into a decimal, and then make that decimal into a percentage. 9/60 = 0.15. We have found our percent as a decimal, and to turn that decimal into a percent, we need to multiply it by 100. Remember that "percent" means "per 100." We know that 0.15 represents "fifteen hundredths," "15 per 100," or "15 percent." Continue to page 15!

ANSWER KEY FOR THIS TYPE OF PROBLEM:

Page 7: Continue to page 15
Page 41 (left): Continue to page 24
Page 41 (right): Continue to page 20
Page 49 (left): Continue to page 25
Page 49 (right): Continue to page 32

MATH CHECKLIST

In your hands is the first ever math *Choose Your Own Adventure* story! This gamebook works differently from an ordinary book. *You* are the main character, and the choices that *you* make on each page will help you uncover just what is going on at Dregg Tower. *What is the Dregg Corporation up to? What will they do if they find you snooping around?*

On each page of your adventure, you will find a math problem, and in order to find the next page in the story, you will need to use your Algebra I skills. First, solve the problem, and then turn to the page that matches your solution. If you solved it correctly, you will find the next page in the story. If you need a little bit of help, this book contains explanations of all the math topics described below.

Here is what *The Dregg Disaster* will help you practice:

Chapter 1: Equations and Proportions

- ❏ Visual equations
- ❏ Solving proportions
- ❏ Using one-step equations
- ❏ Equations with variables on both sides of the equals sign
- ❏ Similar shapes
- ❏ Solving equations with the distributive property
- ❏ Order of operations
- ❏ Solving equations with like terms
- ❏ Proportions with binomial terms
- ❏ Using formulas

Chapter 2: Linear Relationships

- ❏ Using linear patterns
- ❏ Graphing linear inequalities
- ❏ Graphing a system of equations
- ❏ Finding slope from two points
- ❏ Finding y-intercept from two points
- ❏ Solving systems of equations with elimination

- ❑ Graphing lines in slope-intercept form
- ❑ Solving systems of equations with substitution
- ❑ Visual systems of equations

Chapter 3: Quadratic Relationships

- ❑ Quadratic vocabulary
- ❑ Adding polynomials
- ❑ Factoring trinomials
- ❑ Using quadratic patterns
- ❑ Finding solutions with the quadratic formula
- ❑ Solving with square roots
- ❑ Polynomial multiplication
- ❑ Systems of equations with quadratics
- ❑ Finding the vertex of a parabola
- ❑ Factoring with like terms
- ❑ Using the discriminant
- ❑ Subtracting polynomials

Chapter 4: Data Representations and Probability

- ❑ Basic probability
- ❑ Finding IQR from a boxplot
- ❑ Missing values in a two-way frequency table
- ❑ Finding mean from a set of data
- ❑ Finding a missing %
- ❑ Finding median from a set of data
- ❑ Finding % from a two-way frequency table
- ❑ Interpreting a stemplot
- ❑ Finding a new price given % tax or discount
- ❑ Missing value in a set of data, given the mean

Can you take down Dregg? Are your math skills up to the test? Good luck!

INDEX

Page numbers refer to the Adventurer's Advice and not the main text.

Q

R

S

T

ABOUT THE ARTISTS

Illustrator: María Pesado is an illustrator from Barcelona, Spain. After graduating in Graphic Arts and Illustration, she worked as a decorative painter, muralist, and teacher in art workshops. Her paintings have been shown at several exhibitions, and she has directed multidisciplinary events of illustration and scenic arts. In the editorial field, she has published illustrated children's books, and she is currently making her way into comics, collaborating with sci-fi and horror magazines. She lives with her partner, by the sea, and loves books and fantasy movies.

Cover Artist: Eoin Coveney is an Irish illustrator who lives and works in Southern Ireland. After a couple of years in the UK and Germany working as a graphic designer, he returned to Ireland in the mid-1990s. For the last twenty-five years, he has been working with a diverse client base on a wide variety of commercial projects. His aesthetic has been shaped by European comics, horror films, and early twentieth-century illustration. Early in his illustration career, he worked with Will Eisner (renowned comic creator and inventor of the term "Graphic Novel"). From this experience, he gained valuable insight into the process of telling stories through pictures.

ABOUT THE AUTHOR

Chris Matthews is a middle school math teacher from Spokane, Washington, but right now he teaches 7th and 8th grade math in Bucaramanga, Colombia. He loves working with students and creating games that encourage mathematical exploration. When he is not teaching or writing, Chris spends time riding his bike and watching bad movies.

Chris has been teaching math for ten years, and he earned his Masters Degree in Education in 2020. He is also a proud alumnus of Bike & Build, Americorps NCCC, WWU, and Camp Reed. *The Dregg Disaster* is his first book.

ACKNOWLEDGMENTS

This book would not have been possible without the help and support of many people. Thanks foremost to my students and colleagues at MLK, BMS, and CP, who have helped me to grow into the educator that I am today. Thank you for your grace, your patience, your creativity, and your humor. I love teaching because of the people that I get to spend my time with.

Thanks to my wonderful friends and family. Thanks to Mom, Dad, Tyler, Shelby, Barb, Walt, and Emily for your love and support. You encourage me to be creative and honest, and you make my life better each day in ways big and small. You also persevere when I continue to make fart jokes well into thirties. I am very lucky. Thanks, too, to my math brother Chris Bakke, for helping with some of the math in this book and for acting as a sounding board as this idea was taking shape.

Finally, thanks to the team at Chooseco, because this book was quite an undertaking. I originally submitted this story as four enormous PowerPoint files with a bunch of hand-drawn math problems, and Julie, Rachel, Colleen, Peter, and Shannon helped to turn that mess into something worth reading, complete with art (special thanks to María Pesado), color, and energy. I hope you like what this team was able to create.

Want more math?

Get The Dregg Disaster Amazing Free Download! It provides lots of extra practice problems from the algebra you have just completed in this book!

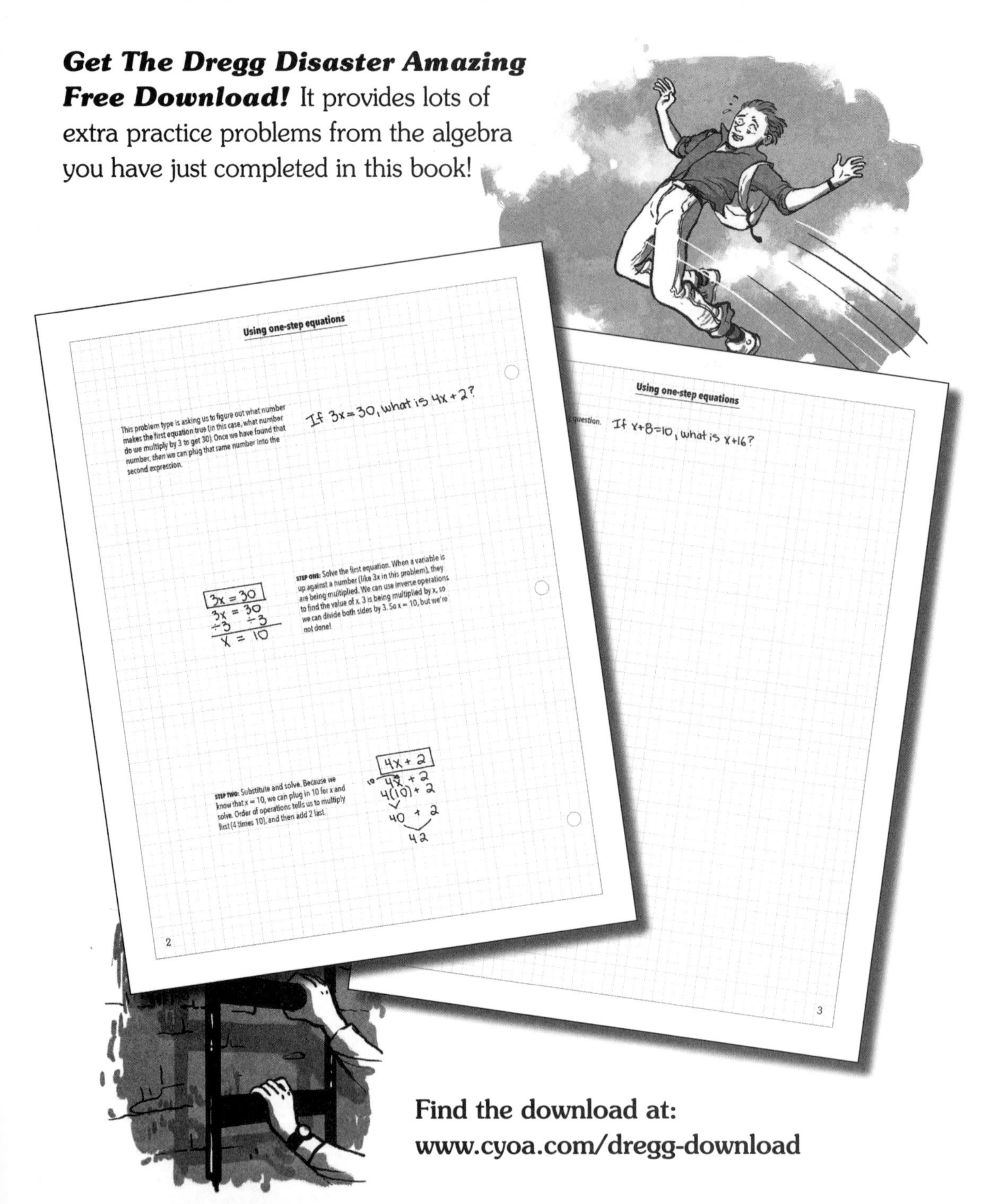

Find the download at:
www.cyoa.com/dregg-download

THE DREGG DISASTER

AMAZING FREE DOWNLOAD

BY CHRIS MATTHEWS

WITH CHRIS BAKKE

TIME TRAVEL Inn

"Mind-blowing adventure and heart-stopping thrills!"

- Jeff Kinney, *Diary of a Wimpy Kid*

"I ended up reading every twisty turn just to make sure I hadn't missed any... arm yourself with your basketball and tiny nunchucks and see where you go!"

- Tina Connolly, Hugo-nominated author of *Seriously Wicked*

MORE PRAISE FOR *TIME TRAVEL INN*

"I was on the edge of my seat with this one! Astrid and her new friends, Damien, Trent, and Davy Cricket, try to solve the mystery of her grandmother's disappearance in this adventure. For fans of King's sense of humor, grab this book of non-stop action and enjoy all the twists and turns it has to offer!"
- Amazon Review

"To have a thrilling, mind-bending adventure, turn to Page 1 of *Time Travel Inn*."
- Dale E. Basye, author of *The Circles of Heck* series

"Travel back in time with Bart King! *Time Travel Inn* takes you across space and time to prehistoric civilizations, pirate ships, and even alternate universes. This fast-paced adventure is literally out of this world!"
- Kate Ristau, folklorist and author of *Clockbreakers*

COMING FALL 2023

THE DREGG DISASTER

This book is different from other *Choose Your Own Adventure* books. It's *certainly* different from other math books. Just like other *Choose Your Own Adventure* books, you will need to be heroic and use your smarts to reach the best ending. Just like other math books, you will need to solve algebra problems as you move from page to page.

Read the story, make your choice, and solve a math problem. You will need to find the correct solution in order to find the next page in your adventure. Just remember: even though there is one correct solution to most of the problems in this book, there are often lots of great ways to find the right answer. Be creative! If you get stuck, there is Adventurer's Advice in the back of each chapter.

A word of caution before you begin your quest: Be careful that the page that you turn to is located in the correct chapter! This book is split into four chapters, and each one has pages numbered 1-50. Don't move on to the next chapter until the book tells you to.

And don't worry. There are still multiple astounding plot twists, unlikely villains, and plenty of death endings.

Good luck!